UNIVERSITY ALGEBRA

THROUGH 600 SOLVED PROBLEMS

UNIVERSITY ALGEBRA

THROUGH 600 SOLVED PROBLEMS

SECOND EDITION

N S GOPALA KRISHNAN

Retired Professor
Department of Mathematics
University of Pune, Pune
Maharashtra

Publishing Globally

NEW AGE INTERNATIONAL (P) LIMITED, PUBLISHERS

LONDON • NEW DELHI • NAIROBI

Bangalore • Chennai • Cochin • Guwahati • Hyderabad • Kolkata • Lucknow • Mumbai

Visit us at **www.newagepublishers.com**

Published by New Age International (P) Ltd., Publishers
First Edition: 1997
Second Edition: 2016
Reprint: 2020

GLOBAL OFFICES

- **New Delhi** — **NEW AGE INTERNATIONAL (P) LIMITED, PUBLISHERS**
  7/30 A, Daryaganj, New Delhi-110002, (INDIA)
  Tel.: (011) 23253771, 23253472, **Telefax:** 23267437, 43551305
  E-mail: contactus@newagepublishers.com • Visit us at www.newagepublishers.com
- **London** — **NEW AGE INTERNATIONAL (UK) LTD.**
  27 Old Gloucester Street, London, WC1N 3AX, UK
  E-mail: info@newacademicscience.co.uk • Visit us at www.newacademicscience.co.uk
- **Nairobi** — **NEW AGE GOLDEN (EAST AFRICA) LTD.**
  Ground Floor, Westlands Arcade, Chiromo Road (Next to Naivas Supermarket)
  Westlands, Nairobi, KENYA, **Tel.:** 00-254-713848772, 00-254-725700286
  E-mail: kenya@newagepublishers.com

BRANCHES

- **Bangalore** — 37/10, 8th Cross (Near Hanuman Temple), Azad Nagar, Chamarajpet, Bangalore- 560 018
  Tel.: (080) 26756823, **Telefax:** 26756820, **E-mail: bangalore@newagepublishers.com**
- **Chennai** — 26, Damodaran Street, T. Nagar, Chennai-600 017, **Tel.:** (044) 24353401, **Telefax:** 24351463
  E-mail: chennai@newagepublishers.com
- **Cochin** — CC-39/1016, Carrier Station Road, Ernakulam South, Cochin-682 016
  Tel.: (0484) 2377303, **Telefax:** 4051304, **E-mail: cochin@newagepublishers.com**
- **Guwahati** — Hemsen Complex, Mohd. Shah Road, Paltan Bazar, Near Starline Hotel, Guwahati-781 008
  Tel.: (0361) 2513881, **Telefax:** 2543669
  E-mail: guwahati@newagepublishers.com
- **Hyderabad** — 105, 1st Floor, Madhiray Kaveri Tower, 3-2-19, Azam Jahi Road, Near Kumar Theater,
  Nimboliadda Kachiguda, Hyderabad-500 027, **Tel.:** (040) 24652456, **Telefax:** 24652457
  E-mail: hyderabad@newagepublishers.com
- **Kolkata** — RDB Chambers (Formerly Lotus Cinema) 106A, 1st Floor, S N Banerjee Road,
  Kolkata-700 014, **Tel.:** (033) 22273773, **Telefax:** 22275247
  E-mail: kolkata@newagepublishers.com
- **Lucknow** — 16-A, Jopling Road, Lucknow-226 001, **Tel.:** (0522) 2209578, 4045297, **Telefax:** 2204098
  E-mail: lucknow@newagepublishers.com
- **Mumbai** — 142C, Victor House, Ground Floor, N.M. Joshi Marg, Lower Parel, Mumbai-400 013
  Tel.: (022) 24927869, **Telefax:** 24915415, **E-mail: mumbai@newagepublishers.com**
- **New Delhi** — 22, Golden House, Daryaganj, New Delhi-110 002, **Tel.:** (011) 23262368, 23262370
  Telefax: 43551305, **E-mail: sales@newagepublishers.com**

978-81-224-3604-4
C-19-08-12203

Printed in India at A.H. Print Pack, New Delhi.
Typeset at Goswami Associates, New Delhi.

NEW AGE INTERNATIONAL (P) LIMITED, PUBLISHERS
7/30 A, Daryaganj, New Delhi-110002
Visit us at **www.newagepublishers.com**
(CIN: U74899DL1966PTC004618)

Preface to the Second Edition

After the good response of the earlier editions it is a pleasure for us to present the second edition. In this edition the style pattern of the book has been modified to give a new look.

In response to the comments made on the First edition by many teachers, presentation of the matter has been improved for the good. The solution and answers of the problems are rechecked and wherever necessary corrections are made. We would like to acknowledge the valuable advice and suggestions of many instructors and students who used the earlier editions. We have tried our best to make the book error free even then some mistakes might have been crept in. Readers are requested to send the same to us for further incorporation in the book.

—Author

Preface to the First Edition

The second and revised edition of **University Algebra,** has been in the field for the last ten years. It has not undergone any revision during this period in spite of best efforts by me and the publishers to revise it. This was mainly due to the fact that there was not enough feedback from the users of the book for revision.

There had been, however, many queries regarding solutions to problems in **University Algebra.** This led to the idea of writing a separate book presenting solutions to the problems in the book.

In this book the solutions are presented together with the original problems so that the book can be used independently as a **Problems/Solutions** book in Algebra rather than just being a "key" to **University Algebra.**

I have avoided the temptation of giving just the hints or giving solutions only to selected problems. Going through all the additional efforts involved I have given complete solutions in a clear and lucid style so that the student can understand in a better way the theory presented in the original book.

The solutions given in the book are not necessarily the shortest or quickest solutions. Our aim is not to present ready-made solutions which can be memorised and reproduced in the examinations but to make the students understand the theory in a better way. The students are expected to try solving these problems themselves and go through the solutions only after some genuine effort. If more students use this book in this manner rather than just using it as a guide for examination, I would consider my effort in writing this book as fruitful.

I would like to take this opportunity to thank Prof. R. Sridharan, Dr. S.A. Katre, A. Acharya and R. Anand for their help in solving certain problems.

Suggestions for improving book are most welcome.

Pune 1997 **N.S. Gopala Krishnan**

Contents

CHAPTER 1

Groups

1.1 BINARY OPERATIONS

1. Let $G = \{1, -1, i, -i\}$ be a set of four elements. Are the following operations binary operations in G? (*i*) $a*b = a + b$ (*ii*) $a*b = a.b$.

Solution: (*i*) No (*ii*) Yes.

2. Define a binary operation on the set of even integers which is different from addition, subtraction or multiplication.

Solution: $a*b = a + b - a.b$.

3. How many binary operations can be defined on a set with two elements?

Solution: The number of maps from $S \times S \to S$ where S has 2 elements is $2^4 = 16$.

4. Define a binary operation on the set of positive integers by $a*b = \text{Max}(a, b)$. Show that this operation is both associative end commutative.

Solution: Associativity follows from $\text{Max}(\text{Max}(a, b), c) = \text{Max}(a, \text{Max}(b, c))$. Commutativity is clear as $\text{Max}(a, b) = \text{Max}(b, a)$.

5. Let $G = \{a, b, c, d\}$ be a set with four elements. Define a binary operation on G by the following table.

	a	*b*	*c*	*d*
a	*a*	*b*	*c*	*d*
b	*b*	*c*	*c*	*d*
c	*c*	*d*	*d*	*a*
d	*d*	*b*	*d*	*c*

Is the binary operation (*i*) Associative, (*ii*) Commutative, (*iii*) Does it have a unit element?

Solution:

(*i*) Not associative as $a = (b * c) * d$ and $b = b * (c * d)$

(*ii*) Not commutative as $b * c \neq c * b$

(*iii*) It has a as the unit element.

6. Define a binary operation on the set of non-negative integers by $m*n = m^2 + n^2$. Does the operation have (*i*) A unit element? (*ii*) An inverse for every element.

Solution: (*i*) No (*ii*) No.

7. Give examples of binary operations on a finite set which is (*i*) Associative but not commutative, (*ii*) Commutative but not associative.

Solution: (*i*) $S = \{a, b\}$

	a	b
a	a	a
b	b	b

is associative but not commutative.

(*ii*) $S = \{a, b, c\}$

	a	b	c
a	a	a	b
b	a	c	a
c	b	a	b

is commutative but not associative as $(ab)\,c \neq a(bc)$.

8. Give an example of a set with two elements and an associative binary operation on it which is not commutative.

Solution: Refer solution to Problem 7(*i*).

9. Show that for the binary operations given below all the following properties are satisfied: (*i*) Associativity, (*ii*) Existence of unit element, (*iii*) Existence of inverse for each element.

(*a*) The set of all real valued continuous functions on $[a, b]$ for the operation $f * g = h$ where $h(t) = f(t) + g(t)$, $0 \le t \le 1$.

(*b*) The set of all ordered pairs of real numbers, for the operation, $(a, b) * (c, d) = (ac - bd, ad + bc)$.

(*c*) The set $G = \{1, -1, i, -i\}$ for the product.

Solution: (*a*) The unit element is the constant function 0 and the inverse of f is $(-f)$.

(*b*) The unit element is $(1, 0)$ and the inverse of $(a, b) = \left(\dfrac{a}{a^2+b^2}, \dfrac{-b}{a^2+b^2}\right)$

(*c*) The unit element is 1 and inverse of 1 is 1, of -1 is -1, of i is $-i$ and of $-i$ is i.

1.2 GROUPS

1. If G is a group, $a, b \in G$. Show that (*i*) $(a^{-1})^{-1} = a$ and (*ii*) $(ab)^{-1} = b^{-1}a^{-1}$.

Solution: (*i*) $aa^{-1} = e = a^{-1}a$ gives $(a^{-1})^{-1} = a$

(*ii*) $(ab)(b^{-1}a^{-1}) = a(bb^{-1})\,a^{-1} = aea^{-1} = aa^{-1} = e$

Also $(b^{-1}a^{-1})(ab) = b^{-1}(a^{-1}a)\,b = b^{-1}eb = b^{-1}b = e$

So $(ab)^{-1} = b^{-1}a^{-1}$.

2. Suppose a group G has an element x such that $ax = x$ for all $a \in G$. Show that G contains only identity element.

Solution: $ax = x$ implies $(ax)\,x^{-1} = xx^{-1} = e$

So $a(xx^{-1}) = e$, and $a = e$. Thus $G = (e)$.

3. Let G be a group, $a, b \in G$. Show that $(aba^{-1})^n = aba^{-1}$ if and only if $b = b^n$.

Solution: $(aba^{-1})\,(aba^{-1}) = ab(a^{-1}\,a)ba^{-1} = ab\,ba^{-1}$

$= ab^2a^{-1}$

By induction $(aba^{-1})^n = ab^na^{-1}$

Thus $(aba^{-1})^n = ab^na^{-1} = aba^{-1}$ if $b^n = b$ and conversely by left and right cancellation.

4. An element $a \in G$ is called idempotent if $a^2 = a$. Show that the only idempotent element in G is the unit element.

Solution: $a^2 = a$ implies $a^{-1}\,(a^2) = a^{-1}a = e$

So $(a^{-1}\,a)a = ea = a = e.$

5. Find a solution of the equation $ax = b$ in S_3, where

$$a = \begin{pmatrix} 1 & 2 & 3 \\ 3 & 2 & 1 \end{pmatrix} \text{ and } b = \begin{pmatrix} 1 & 2 & 3 \\ 1 & 3 & 2 \end{pmatrix}$$

Solution: Solution of $ax = b$ is given by $x = a^{-1}b$.

So $$x = \begin{pmatrix} 1 & 2 & 3 \\ 3 & 2 & 1 \end{pmatrix}^{-1} \begin{pmatrix} 1 & 2 & 3 \\ 1 & 3 & 2 \end{pmatrix} = \begin{pmatrix} 1 & 2 & 3 \\ 3 & 2 & 1 \end{pmatrix}\begin{pmatrix} 1 & 2 & 3 \\ 1 & 3 & 2 \end{pmatrix} = \begin{pmatrix} 1 & 2 & 3 \\ 3 & 1 & 2 \end{pmatrix}$$

6. Show that the set of all transformations of the type $z \to \dfrac{az+b}{cz+d}$, $ad - bc \neq 0$, of complex numbers to itself is a group for composition of transformation.

Solution: Consider $$z \to \frac{az+b}{cz+d},\ ad - bc \neq 0$$

and $$z \to \frac{a'z+b'}{c'z+d'},\ a'd' - b'c' \neq 0$$

Consider composition of the two transformations *viz.*,

$$z \to \frac{a'\left(\dfrac{az+b}{cz+d}\right)+b'}{c'\left(\dfrac{az+b}{cz+d}\right)+d'} = \frac{Az+B}{Cz+D} \text{ (say)}$$

where $AD - BC = (ad - bc)\,(a'd' - b'c') \neq 0$, by simplification.

Thus the operation is closed. Identify corresponds to $a = 1$, $b = 0$, $c = 0$ and $d = 1$.

The inverse of $z \to \dfrac{az+b}{cz+d}$ can be solved from the equations $A = 1$, $B = 0$, $C = 0$, $D = 1$ and there is a unique solution.

7. Show that the six transformations of the type $t \to t, t \to \frac{1}{t}, t \to 1-t, t \to \frac{1}{1-t}$, $t \to \frac{t-1}{t}$ and $t \to \frac{t}{t-1}$ form a group, for the composition of transformations.

Solution: Closure can be verified directly. The identity is $t \to t$ and each element has an inverse, namely the succeeding element.

8. If G is a group such that $a^2 = e$ for every $a \in G$, show that G is abelian. Is the same true if $a^3 = e$ for all $a \in G$?

Solution: $a^2 = e$ implies $(ab)^2 = e$

$$(ab)(ab) = e \text{ and so } a^{-1}(ab)(ab)b^{-1} = a^{-1}eb^{-1}$$

Thus $ba = a^{-1}b^{-1} = ab$. Hence G is abelian.

If $a^3 = e$ for all $a \in G$, G need not be abelian.

9. Show that G is an abelian group, if and only if $(ab)^2 = a^2b^2$, for all $a, b \in G$.

Solution: If G is abelian

$$(ab)^2 = (ab)(ab)$$
$$= a(ba)b = a(ab)b = a^2b^2$$

Conversely if $(ab)(ab) = a^2b^2$ then

$$a^{-1}(ab)(ab)b^{-1} = a^{-1}(a^2b^2)b^{-1}$$

and so $ba = (a^{-1}a^2)(b^2b^{-1}) = ab$

Hence G is abelian.

10. Let G be a finite group with even number of elements. Show that there is at least one $a \in G, a \neq e$ such that $a^2 = e$.

Solution: Suppose $a \neq a^{-1}$ for each $a \in G, a \neq e$, pairing them together and along with e, G will have odd number of elements, a contradiction.

11. Show that for any set S, $G(S)$, the set of all (1, 1) mappings of S onto S is a group for composition.

Solution: Composite of two (1, 1) "onto" maps is also (1, 1) and "onto". Inverse of (1, 1) "onto" map is also (1, 1) and "onto".

12. Let $D_n = \{e, a, a^2, \ldots, a^{n-1}, b, ba, \ldots, ba^{n-1}\}$ be a set of $2n$ elements. Define the product in D_n by the relations $a^n = e$. $b^2 = e$ and $ab = ba^{-1}$. Show that these relations can be used to define the product of any two elements in D_n and that D_n is a group. (nth Dihedral group).

Solution: $a^2b = a(ab) = a(ba^{-1}) = (ab)a^{-1}$

$$= (ba^{-1})a^{-1} = ba^{-2}$$

Similarly, $a^ib = ba^{-i}$

Also $(ba^i)(ba^j) = b(ba^{-i})a^j = a^{-i+j}$

Thus the product of any two elements is defined, e is the identity and

$$(a^i)^{-1} = a^{n-i}, b^{-1} = b \text{ and } (ba^j)^{-1} = a^{n-j}b.$$

Hence D_n is a group.

13. Interpret D_n as a group of symmetries of a regular n-sided polygon.

Solution: Let a be the rotation of the polygon through the centre by an angle $2\pi/n$ and b the reflection through the line l where l is the line through a vertex perpendicular to the opposite side (n odd) or l is the line perpendicular to a pair of opposite sides (n even). The elements a and b satisfy the relations in the previous problem defining D_n. Thus D_n is a group of symmetries of a regular n sided polygon.

14. Show that the set of all rotations in the 3-space given by $(x, y, z) \to (x', y', z')$ is a group for composition where

$$x' = a_1x + b_1y + c_1z$$
$$y' = a_2x + b_2y + c_2z$$
$$z' = a_3x + b_3y + c_3z$$

with $\quad a_i^2 + b_i^2 + c_i^2 = 1, (1 \le i \le 3)$ and $a_ia_j + b_ib_j + c_ic_j = 0, i \ne j$

and $\quad \begin{vmatrix} a_1 & b_1 & c_1 \\ a_2 & b_2 & c_2 \\ a_3 & b_3 & c_3 \end{vmatrix} = 1.$

Solution: The transformation is $\begin{pmatrix} x' \\ y' \\ z' \end{pmatrix} = A \begin{pmatrix} x \\ y \\ z \end{pmatrix}$

where $A = \begin{pmatrix} a_1 & b_1 & c_1 \\ a_2 & b_2 & c_2 \\ a_3 & b_3 & c_3 \end{pmatrix}$ A satisfies $A^tA = AA^t = I$ and $|A| = 1$. If A, B satisfies these relations so does AB and A^{-1}. Identity map clearly satisfies it. Hence, these transformations form a group.

1.3 SUBGROUPS

1. Let $G = C^*$ be the group of non-zero complex numbers for multiplication. Are the following subgroups of G?

(*i*) All purely imaginary complex numbers.

(*ii*) All complex numbers with absolute value 2.

Solution: (*i*) No (*ii*) No.

2. Show that the set S of all transformations of the type $z \to \dfrac{az+b}{cz+d}$, $ad - bc = 1$, is a sub-group of the group defined in Problem 6, Section 1.2.

Solution: The composite of $z \to \dfrac{az+b}{cz+d}$ and $z' \to \dfrac{a'z+b'}{c'z+d'}$ is $z \to \dfrac{Az+B}{Cz+D}$, where $AD - BC = (ad - bc)(a'd' - b'c') = 1$.

Hence S is closed for the group operation. Also the inverse of $z \to \dfrac{az+d}{cz+d}$ satisfies the condition $a'd' - b'c' = 1$. Hence S is a subgroup.

3. Give an example to show that union of two subgroups may not be a subgroup.

Solution: $G = Z$, $H_1 = 2Z$, $H_2 = 3Z$. Then $H = H_1 \cup H_2$ is not a subgroup as $2 \in H$, $3 \in H$ and hence 1 should be in H *i.e.* H should be equal to G. But $H \neq G$ as $5 \in H$ and $5 \notin H$.

4. If K is a subgroup of H and H is a subgroup of G, show that K is a subgroup of G.

Solution: If, $a, b \in K$, then $ab^{-1} \in K \subset G$.

5. If G is an abelian group, show that $H = \{a \mid a \in G, a^2 = e\}$ is a subgroup of G.

Solution: If $a, b \in H$, than $(ab)^2 = a^2 b^2$ (G abelian) $= e.e = e$. Hence $ab \in H$. Also $e \in H$ and $a^{-1} \in H$ if $a \in H$. Thus H is a subgroup.

6. Find all subgroups of (*i*) D_3 (*ii*) D_4 (*iii*) D_n.

Solution: (*iii*) $D_n = (e, a, a^2, ..., a^{n-1}, b, ba, ..., ba^{n-1})$ with $a^n = e$, $b^2 = e$ and $ab = ba^{-1}$

H_1 (a) and $H_2 = (b)$ are cyclic subgroups of D_n.

Any other subgroup of D_n is either a subgroup of (a). *i.e.* of the type (a^i) or a subgroup with elements

$$\{e, a^i, a^{2i}, ..., a^{(n/i-1)i}, b, ba^i, ..., ba^{(n/i-1)i}\}.$$

(*i*) and (*ii*) are special cases of D_n for $n = 3, 4$.

7. Let H be subgroup of G and $a \in G$. Show that $a \in H$ if and only if $aH = H$.

Solution: If $a \in H$, $aH \subset H$. Also $H \subset aH$ because if $h \in H$ $h = aa^{-1} h \in aH$, so $aH = H$.

Conversely if $aH = H$, then $a = ae \in H$.

8. If $a \in G$, show that the centraliser $N(a)$ of a is a subgroup of G and $Z(G) \subset N(a)$.

Solution: $N(a) = \{x \mid xa = ax\}$. Clearly $x, y \in N(a)$ implies $xy \in N(a)$ as

$$a(xy) = (ax)\,y = (xa)\,y = x(ay) = x(ya) = (xy)\,a$$

and $x^{-1} \in N(a)$ as $xa = ax$ implies by multiplying with x^{-1} on both left and right $x^{-1} a = ax^{-1}$. Clearly $e \in N(a)$.

Also $Z(G) \subset N(a)$ from the definition of $Z(G)$.

9. Show that for any subset A of G, the normaliser $N(A)$ of A is a subgroup of G. If H is a subgroup of G show that $H \subset N(H)$.

Solution: $N(A) = \{x \mid xA = Ax\}$

If $x, y \subset N(A)$, a proof similar to the one given in the previous exercise shows that $xy \in N(A)$, $e \in N(A)$ and $x^{-1} \in N(A)$. Hence $N(A)$ is a subgroup.

If H is a subgroup and $x \in H$, $xH = H = Hx$ so that $x \in N(H)$.

10. Show that a group cannot be expressed as a union of two proper subgroups.

Solution: Let $G = H \cup K$ be a proper union.

Since $H \not\subset K$ and $K \not\subset H$, choose $a \in H$, $a \notin K$ and $b \in K$, $b \notin H$. Then $ab \in G$ and $ab \notin H$ as well as $ab \notin K$, a contradiction.

11. If G has no non-trivial subgroups, show that G is of prime order.

Solution: Choose $a \in G$, $a \neq e$. Then $(a) \subset G$ and hence $(a) = G$, *i.e.* G is cyclic. Further G has to be finite cyclic for otherwise it will have proper subgroups. Moreover $o(G) = n$ should be prime otherwise, there exists a proper divisor m of n and $(a^{n/m})$ is a proper subgroup of G.

12. Find a non-trivial proper subgroup of the group defined in Exercise 14, Section 1.2.

Solution: Consider all transformation with additional restrictions,

$$a_1 = 1, b_1 = 0, c_1 = 0, a_2 = 0, a_3 = 0.$$

These correspond to matrices $\begin{pmatrix} 1 & 0 & 0 \\ 0 & b_2 & c_2 \\ 0 & b_3 & c_3 \end{pmatrix} = A$ and they form a subgroup.

1.4 CYCLIC GROUPS

1. Are the following groups cyclic? Why?

(*i*) Klein-4-group.

(*ii*) D_4

(*iii*) The group of roots (real or complex) of the equation $x^n - 1 = 0$.

(*iv*) The group Q^* of non-zero retionals, for multiplication.

Solution: (*i*) No, as there is no element of order 4.

(*ii*) No, as it is not abelian.

(*iii*) Yes, the generator is $a = e^{2\pi i/n}$

(*iv*) No, for if $Q^* = (a)$ with $a = p/q$, then $p/q + 1 \in Q^*$ but not in (a).

2. Give an example of a group which is not cyclic, but every proper subgroup of which is cyclic.

Solution: The group S_3.

3. Show that a group which has only finitely many subgroups is a finite group.

Solution: $G = \bigcup_{a \in G} (a)$ and this union is a finite union as G has only finitely many subgroups. Moreover each (a) is finite for otherwise it will have infinitely many subgroups. Hence G is a finite group.

4. Let G be a group such that $G = \bigcup_{n=1}^{\infty} G_n$, $G_n \not\subseteq G_{n+1}$. Show that G cannot be cyclic.

Solution: If G is cyclic, $G = (a)$ and $a \in G_n$ for some n. Then $(a) \subset G_n \subset G_{n+1} \subset ... \subset G = (a)$ i.e. $G_n = G_{n+1} = ...$ a contradiction.

5. Show that a cyclic group with one generator has utmost 2 elements.

Solution: Let $G = (a)$ with one generator.

Then $o(G) < \infty$ and if $o(G) = n$, then the number of generators is $\phi(n) = 1$. Hence $n = 2$.

6. Let $a, b \in G$ such that $b = xax^{-1}$, for some $x \in G$. Show that $o(a) = o(b)$.

Solution: $b = xax^{-1}$. By iteration $b^n = xa^n x^{-1}$.

So $b^n = e$ if, and only if, $a^n = e$. Thus $o(a) = o(b)$.

7. Let $a, b \in G$ be any two elements. Show that $o(ab) = o(ba)$.

Solution: $ab = b^{-1}(ba)\, b$ and by the previous exercise $o(ab) = o(ba)$.

8. Let $a \in G$ be an element of order 2 and $b \in G$ an element of order 3. Show that if $H = (a)$ and $K = (b)$ and $ab = ba$, then HK is a group of order 6.

Solution: Consider $HK = \{e, a, b, b^2, ab, ab^2\}$. It is a subgroup of G because

$$HK = KH \ (ab = ba).$$

Also
$$o(HK) = \frac{o(H)\cdot o(K)}{o(H \cap K)} = \frac{2\cdot 3}{1} = 6.$$

9. If G is an abelian group, show that all elements of finite order in G form a subgroup of G (called torsion subgroup of G).

Solution: Let $T = \{a \mid a \in G, o(a) < \infty\}$.

If $a, b \in T$, then $a^n = e = b^m$, $(ab)^{mn} = a^{mn} \cdot b^{mn} = e$ and so $o(ab) < \infty$.

Also $o(a^{-1}) = o(a) < \infty$. Since $e \in T$, T is a subgroup.

10. Let $a \in G$ be such that $o(a) = mn$. Show that if $b = a^m$, then $o(b) = n$.

Solution: $b^n = a^{mn} = e$. Also $b^i = a^{im} \neq e$ if $i < n$ as $o(a) = mn$. Thus $o(b) = n$.

11. Let G be a cyclic group of order p^n ,where p is a prime, H and K subgroups of G. Show that either $H \subset K$ or $K \subset H$.

Solution: Let $G = (a)$, $o(a) = p^n$. H is cyclic and so $H = (a^i)$ with $i \mid p^n$, i.e. $i = p^\lambda$.

Similarly $K = (a^i)$ with $j = p^\mu$.

If $\lambda \geq \mu$, $H \subset K$, otherwise $K \subset H$.

1.5 LAGRANGE'S THEOREM

1. Let $G = Z$ and $H = mZ$. What is $(G : H)$?

Solution: $(G : H) = m$.

2. Let $G = (R, +)$, and $H = (2\pi)$. Show that the sine function assigns the same value to each element of any coset of G modulo H.

   Solution: A typical coset is of the type $x + H$ and $\sin(x + H) = \sin x$ as elements of H are integral multiples of 2π.

3. Let G be a group, H a subgroup of G such that $(G : H) = 2$. Show that every left coset of G mod H is also a right coset.

   Solution: Since $(G : H) = 2$, the set of left cosets is $\{H, aH\}$, $a \notin H$ and the set of right cosets is $\{H, Ha\}$. Clearly $Ha = G - H = aH$.

4. Let G be a non-trivial group which has no proper subgroups. Show that G is a cyclic of prime order.

   Solution: Choose $a \in G$, $a \neq e$. Then $(a) = G$ and hence G is cyclic. Let $o(G) = n$.

   If m is a proper divisor of n then $(a^m) \not\subseteq G$ a contradiction. Hence n is prime.

5. Let G be a group, H and K subgroups of G such that H is a finite group of prime order, and $H \cap K \neq \{e\}$. Show that $H \subset K$.

   Solution: $o(H) = p$, prime and $o(H \cap K) \neq 1$

   and hence $H = H \cap K$. So $H \subset K$.

6. Let G be a finite group, H, K subgroups of G such that $K \subset H$.

   Show that $(G : K) = (G : H)\,(H : K)$.

   Solution: $G = \cup\, aH$ a disjoint union of m cosets

   and $H = \cup\, bK$ a disjoint union of n cosets

   where $(G : H) = m$ and $(H : K) = n$.

   Then $G = \cup\, abK$ is a disjoint union of mn cosets

   Thus $(G : K) = mn = (G : H)\,(H : K)$.

7. Let G be a group, H and K finite subgroups of G such that $o(H)$ and $o(K)$ are relatively prime. Show that $H \cap K = \{e\}$.

   Solution: Let $o(H) = m$ and $o(K) = n$ so that $(m, n) = 1$.

   $$o(H \cap K) \mid o(H) = m \quad \text{and} \quad o(H \cap K) \mid o(K) = n.$$

   So $o(H \cap K) \mid (m, n) = 1$. Thus $H \cap K = \{e\}$.

8. Let G be a group, H and K subgroups of finite indices in G. Show that $H \cap K$ is also of finite index. If $(G : H) = m$ and $(G : K) = n$ with $(m, n) = 1$, show that $(G : H \cap K) = mn$.

   Solution: $(G : H \cap K) = (G : H)\,(H : H \cap K)$.

   Since $(G : K) < \infty$ it follows that $(H : H \cap K) \le (G : K) < \infty$

   and so $(G : H \cap K) \le mn < \infty$

   Now $(G : H) \mid (G : H \cap K)$ and $(G : K) \mid (G : H \cap K)$

   So that their product $mn \mid (G : H \cap K)$ if $(m, n) = 1$

   and so $mn \le (G : H \cap K)$. Hence $mn = (G : H \cap K)$.

9. If G is a cyclic group of order n, show that G has exactly $d(n)$ subgroups, where $d(n)$ is the number of divisors of n.

Solution: Let $G = (a)$ be of order n.

For each $d \mid n$, $(a^{n/d})$ is a subgroup of order d and it is a unique subgroup of order d for if (a^i) is a subgroup of order d, $id = n$, i.e. $i = n/d$. Conversely every subgroup of G has order d, a divisor of n. Thus the number of distinct subgroups of G is equal to the number of divisors of n.

10. Using the foregoing exercise, show that for any positive integer n, $n = \sum_{d/n} \phi(d)$.

Solution: Consider a cyclic group G order n. Each subgroup of order d had $\phi(d)$ generators and counting all these elements exhaust all the elements of G as each elements of G is a generator of some subgroup. Hence $\sum_{d/n} \phi(d) = n$.

11. Define the exponent of a group as $\sup_{a \in G} o(a)$. Show that if G is a finite abelian group as $o(a)$ divides the exponent of G for each $a \in G$.

Solution: For any $a \in G$, let $o(a) = n$.

If m = exponent of G, to show that $n \mid m$. Suppose this is not true. Write $n = \prod p_i^{\alpha_i}$ and $m = \prod p_i^{\beta_i}$. Then at least for some i, $\alpha_i > \beta_i$. Choose an element $b \in G$ with $o(b) = m$. Consider the elements

$$a' = a^{n/p_i^{\alpha_i}} \quad \text{and} \quad b' = b^{p_i^{\beta_i}}.$$

Then $o(a') = p_i^{\alpha_i}$ and $o(b') = \prod_{j \neq i} p_j^{\beta_j}$ and these are relatively prime. Since G is abelian $o(a'b') = p_i^{\alpha_i} \prod_{j \neq i} p_j^{\beta_j} > m$, as $\alpha_i > \beta_i$. This is a contradiction to the definition of exponent.

12. Let H be a subgroup of finite index. Show that there exist $x_1, x_2, ..., x_r \in G$, $r = (G : H)$ such that $G = \bigcup_{i=1}^{r} x_i H = \bigcup_{i=1}^{r} x_i H$.

Solution: We first observe that H contains a normal subgroup N which is also of finite index in G. To show this, we note that H has only finite conjugates as $aHa^{-1} = aH \cdot Ha^{-1}$ and $(G : H) < \infty$. Let these be $H_1, H_2, ..., H_m$. Then $N = \bigcup_{i=1}^{m} H_i$ is normal in G and also has finite index by Exercise 8.

Since N is normal, there exists $x_1, ..., x_r$ in G such that

$G = \cup x_i N = \cup N x_i$. Since H is a conjugate of H, $N \subset H$

So, $G = \cup x_1 H = \cup H x_i$.

13. Let G be a finite group of order n. Show that G is cyclic if, and only if, n is the smallest positive integer such that $a^n = e$, for some $a \in G$.

Solution: If $o(G) = n$ and $G = (a)$ is cyclic, clearly $a^n = e$, $a^m \neq e$ for $m < n$. Conversely assume that n is the smallest positive integer such that $a^n = e$ for some $a \in G$. G has an element a of order $n = o(G)$. Hence G is cyclic.

1.6 ISOMORPHISM

1. Among the following groups, find pairs of groups which are isomorphic.
   (*i*) The group Z of integers.
   (*ii*) The additive group R of real numbers.
   (*iii*) The additive group of all integral multiples of π.
   (*iv*) The group of residue classes modulo 4.
   (*v*) The multiplicative group of positive real numbers.
   (*vi*) The group $G = \{1, -1, i, -i\}$.
   Solution: (*i*) and (*iii*) are isomorphic.
   (*ii*) and (*v*) are isomorphic under the map $x \to e^x$
   (*iv*) and (*vi*) are isomorphic, both cyclic of order 4.
2. Show that the additive group Z_4 is isomorphic to the multiplicative group of non-zero elements of Z_5.
   Solution: Both are cyclic groups of order 4.
3. Show that if G is isomorphic to G', $Z(G)$ is isomorphic to $Z(G')$.
   Solution: Let $f : G \to G'$ be an isomorphism to Z onto G'. Then $x \in Z(G)$ if, and only if, $x' = f(x) \in Z(G')$ as $ax = xa$, for all $a \in G$ implies $a'\,x' = x'\,a'$ for all $a' \in G'$. Thus f restricted to $Z(G)$ is an isomorphism of $Z(G)$ onto $Z(G')$.
4. Find all automorphisms of Z_6.
   Solution: Z_6 is a cyclic group of order 6.
   Let a be a generator. Then $a \to ia$ is an automorphism if, and only if, $(i, 6) = 1$. Thus $i = 1, 5$. $i = 1$ corresponds to Id and $i = 5$ corresponds to an automorphism
   $\varphi : Z_6 \to Z_6$, where $\phi\ (\overline{0}) = 0$, $\phi\ (\overline{1}) = \overline{5}$, $\phi\ (\overline{2}) = \overline{4}$, $\phi\ (\overline{3}) = \overline{3}$, $\phi\ (\overline{4}) = \overline{2}$ and $\phi\ (\overline{5}) = \overline{1}$.
5. Show that the set of all automorphisms of G form a group Aut (G) for composition.
   Solution: Composite of two automorphism of G onto G is an automorphism, $Id \in$ Aut G and inverse of an automorphism is an automorphism.
6. Give an example of a non-trivial group which has only trivial automorphisms.
   Solution: Let G be a cyclic group of order 2. Then Aut $(G) = (I)$.
7. Show that the set $I(G)$ of inner automorphisms of G is a subgroup of the Aut (G).
   Solution: Let $f_a : G \to G$ be the inner automorphism defined by $f_a\,(x) = axa^{-1}$.
   Then $$f_b f_a\,(x) = b\,(axa^{-1})\,b^{-1} = (ba)\,x\,(ba)^{-1} = f_{ba}\,(x)$$

Thus $\qquad f_b f_a = f_{ba}$. Also $f_e = Id$ and $f_a^{-1} = f_a^{-1}$.

Hence $I(G)$ is a subgroup of Aut (G).

8. If θ is an automorphism of G, show that for any $a \in G$ the elements a and $\theta(a)$ have the same orders.

Solution: $\theta : G \to G$ is an automorphism. Then $a^n = e$ if, and only if, $\theta(a^n) = \theta(a^n) = e$. Hence $o(a) = o(\theta\,(a))$.

9. If G is the Klein-4-group, show that Aut $(G) \approx S_3$.

Solution: $G = \{e, a, b, c = ab\}$ with $a^2 = e = b^2$ and $ab = ba$.

Let $\sigma \in$ Aut (G). σ is determined by its values on a, b and c and these would be any of a, b or c but distinct.

Thus σ is a permutation on 3 symbols a, b c and Aut $(G) \approx S_3$.

10. Find all the automorphisms of S_3.

Solution: Each automorphism of S_3 permutes $a = (1\ 2)$, $b = (2\ 3)$ and $c = (1\ 3)$ and no nontrivial automorphism leaves each element fixed. Hence if two automorphisms σ and τ permute the set in the same way $\sigma\tau^{-1} = Id$, *i.e.* $\sigma = \tau$.

Thus S_3 has utmost $\underline{|3} = 6$ automorphisms.

However $I(S_3) \approx \dfrac{S_3}{Z(S_3)} \approx S_3$ by Exercise 4, of 1.8 as $Z(S_3) = \{e\}$ and has 6 elements.

So $\qquad I(S_3) = \text{Aut}\ (S_3) \approx S_3.$

11. Let G be a finite group and $\theta : G \to G$ an automorphism such that $\theta^2 = Id$ and $\theta(a) = a$ implies $a = e$. Show that G is abelian.

Solution: We first note that $G = \{a^{-1}\,\theta\,(a) \mid a \in G\}$.

Suppose $\qquad a^{-1}\,\theta\,(a) = b^{-1}\,\theta\,(b).$

Then $\qquad ba^{-1} = \theta\,(b)\,\theta\,(a)^{-1} = \theta\,(ba^{-1})$

and by assumption on θ, $ba^{-1} = e$, *i.e.* $b = a$.

Thus all the elements $a^{-1}\,\theta\,(a)$ are distinct and since G is finite

$$G = \{a^{-1}\,\theta\,(a) \mid a \in G\}.$$

We claim now that $\theta\,(x) = x^{-1}$ for any $x \in G$. Write $x = a^{-1}\,\theta\,(a)$ for some $a \in G$.

$$\theta\,(x) = \theta\,(a)^{-1}\,\theta^2\,(a) = \theta\,(a^{-1}) \cdot a = (a^{-1}\,\theta\,(a))^{-1} = x^{-1}.$$

Thus $x \to x^{-1}$ is an automorphism of G. This implies that G is abelian as

$$\theta\,(xy) = (xy)^{-1} = y^{-1}\,x^{-1} = \theta\,(y)\,\theta\,(x) = \theta\,(yx) \text{ i.e. } xy = yx.$$

12. Let G be a finite cyclic group of order n. Show that Aut (G) is isomorphic to the group $U\,(Z_n)$ of prime residue classes modulo n.

Solution: Let $G = (a)$, with $o(a) = n$.

Let $\theta : G \to G$, be $\theta\,(a) = a^i$. Then θ is an automorphism if, and only if, $(i, n) = 1$. The map

$$\text{Aut } (G) \to U(Z_n)$$

$$\theta \to \bar{i} \text{ is an isomorphism of groups.}$$

13. Let G be a group such that Aut $(G) = \{Id\}$. Show that G is abelian and that $a^2 = e$ for all $a \in G$. If, further, G is finite and cyclic show that $o(G) = 1$ or 2.

Solution: Since Aut $(G) = \{Id\}$, $I(G) = \{I\}$.

i.e. $\quad f_a = Id$ for all $a \in G$.

i.e. $\quad f_a(x) = x$ for all x an so $ax = xa$ for all $x \in G$.

Hence G is abelian. Since $a \to a^{-1}$ is an automorphism, it has to be Id, *i.e.* $a^{-1} = a$ for all a, *i.e.* $a^2 = e$ for all $a \in G$.

It G is finite cyclic of order n, the number of automorphism of G is $\phi(n)$. $\phi(n) = 1$ implies $n = 1$ or 2.

14. True or false? Justify: G is isomorphic to a subgroup of G' and G' is isomorphic to a subgroup of G, then G is isomorphic to G'.

Solution: Not True. Consider the Example. Let G be the free (non-abelian) group on 2 generators x and y. Let G' be the free subgroup of G generated by x^2, xy and y^2. The G is isomorphic to the subgroup H of G', generated by x^2 and y^2 as both are free groups on 2 generators. Also G' is a subgroup of G. But G is not isomorphic to G' as G' is free on 3 generators.

1.7 HOMOMORPHISM

1. Let R^* be the multiplicative group of non-zero reals. Which of the following mappings of R^* in R^* are group homomorphisms? In case they are group homomorphisms find the kernel.

(*i*) $x \to x^2$ $\qquad$ (*ii*) $x \to |x|$ $\qquad$ (*iii*) $x \to \dfrac{1}{x}$.

Solution: (*i*) Yes. Ker $= \{\pm 1\}$

(*ii*) Yes. Ker $= \{\pm 1\}$

(*iii*) Yes. Ker $= \{1\}$.

2. Let $G = \{z \mid |z| = 1\}$ be the multiplicative group of complex numbers of absolute value 1, and m a positive integer. Show that the mapping $\theta : G \to G$ defined by $\theta(z) = z^m$ is a homomorphism of G onto G. What is the kernel of θ?

Solution: $\quad \theta(zz') = (zz')^m = z^m z'^m = \theta(z) \cdot \theta(z')$.

θ is onto because for each $z \in G$, there exists a ω such that $\omega^m = z$ and $|\omega|^m = |z| = 1$ and so $|\omega| = 1$ *i.e.* $\omega \in G$. Now Ker θ is the set of m-th roots of unity.

3. Let $G = S_3$ and $G' = Z_3$. Show that there is no homomorphism of G in G' except the trivial homomorphism.

Solution: Let $\theta : S_3 \to Z_3$ be a homomorphism. Ker θ is a normal subgroup of S_3. If Ker $\theta = \{e\}$, S_3 will be isomorphic to a subgroup of Z_3 which is impossible. If Ker $\theta = A_3$, A_3/S_3 is isomorphic to a subgroup of Z_3 which is impossible as it has order 2. Hence Ker $\theta = S_3$, *i.e.* θ is trivial.

4. Give an example of a subgroup of S_3 which is not normal in S_3.

Solution: Let $\phi = (1, 2)$. Then $\{e, \phi\}$ is a subgroup and it is not normal.

5. If K is normal subgroup of G and H is any subgroup, show that $H \cap K$ is normal in H.

Solution: Clearly $H \cap K$ is a subgroup of H. If $a \in H \cap K$ and $x \in H$, $xax^{-1} \in K$ as K is normal in G and it is also in H. So $xax^{-1} \in H \cap K$ *i.e.* $H \cap K$ is normal in H.

6. Let G be a finite group and H the only subgroup of order $o(H)$ in G. Show that H is normal in G.

Solution: Consider any conjugate xHx^{-1} of H. It is a subgroup of order $o(H)$ and so $xHx^{-1} = H$. Hence H is normal in G.

7. Show that a subgroup H is normal in G if, and only if, the product of two left cosets of H in G is a left coset of H in G.

Solution: If H is normal.

$$(aH)\ (bH) = a\ (Hb)\ H = a\ (bH)\ H = (ab)\ HH = abH$$

Conversely if $(aH)\ (bH)$ is a left coset, it must be a left coset containing ab and so $(aH)\ (bH) = abH$. In particular,

$$(aH)\ (a^{-1} H) = aa^{-1}\ H = H \text{ i.e. } (aHa^{-1})\ H = H$$

and so $aHa^{-1} = HH^{-1} = H$. Thus H is normal.

8. Give an example of a non-abelian group, all of whose subgroups are normal.

Solution: $G = \{\pm 1, \pm i, \pm j, \pm k\}$ with multiplication

$$i^2 = j^2 = k^2 = -1,\ ij = -ji = k,\ jk = -kj = i,\ ki = -ik = j.$$

G is a non-abelian group with all subgroups normal in G.

9. Let $\theta : G \to G'$ be a homomorphism of G in G' with G' abelian. Let H be a subgroup of G containing Ker θ. Show that H is normal in G.

Solution: Let $h \in H$ and $a \in G$. Then

$$\theta\ (aha^{-1}) = \theta\ (a)\ \theta\ (h)\ \theta\ (a)^{-1} = \theta\ (a)\ \theta\ (a)^{-1}\ \theta\ (h)$$

(as G' is abelian)

$$= \theta(h)$$

Thus $\theta\ (aha^{-1}\ h^{-1}) = \theta\ (a)\ \theta\ (h)\ \theta\ (a^{-1})\ \theta\ (h)^{-1} = e$

So $aha^{-1}\ h^{-1} \in \text{Ker}\ \theta \subset H$

Hence $aha^{-1} \in H$ and so H is normal.

10. Let $\theta : G \to G'$ be a homomorphism, with G a simple group. Show that either θ is trivial or a (1, 1) mapping.

Solution: Ker θ is a normal subgroup of G and since G is simple Ker $\theta = G$ or Ker $\theta = \{e\}$. In the former case θ is trivial and in the later case it is (1, 1).

11. Show that a simple abelian group is cyclic of prime order.

Solution: Let G be simple abelian. If $a \in G$, $a \neq e$, (a) is a normal subgroup of a and hence $(a) = G$ *i.e.* G is cyclic.

If $o(G) = n$ is not prime, it has a proper divisor d and there is a proper subgroup of order d which is also normal, a contradiction. Hence $o(G)$ must be prime.

12. Show that the group $I(G)$ of inner automorphism of G is a normal subgroup of Aut (G).

Solution: Let $f_a \in I(G)$ and $\sigma \in \text{Aut}(G)$

Then
$$\begin{aligned}\sigma^{-1} f_a \sigma(x) &= \sigma^{-1}(a\,\sigma(x)\,a^{-1}) \\ &= \sigma^{-1}(a)\cdot x\,\sigma^{-1}(a^{-1}) \\ &= f^{-1}_{\sigma(a)}(x)\end{aligned}$$

Thus $\sigma^{-1} f_a \sigma = f^{-1}_{\sigma(a)} \in I(G)$ and $I(G)$ is normal.

13. Show that any normal subgroup of G of order 2 is contained in $Z(G)$.

Solution: Let $H = \{e, a\}$ be normal of order 2. For any $x \in G$, $xax^{-1} \in H$ and so $xax^{-1} = a$, *i.e.* $xa = ax$. Thus $a \in Z(G)$ and $H \subset Z(G)$.

1.8 QUOTIENT GROUPS

1. Show that it is impossible to find a homomorphism of Z_5 onto Z_4.

Solution: If $\theta : Z_5 \to Z_4$ is a homomorphism "onto".

$\dfrac{Z_5}{\text{Ker}\,\theta} \approx Z_4$. In particular $o\left(\dfrac{Z_5}{\text{Ker}\,\theta}\right) = o(Z_4)$ which is impossible as $4 \nmid 5$.

2. Write down all quotient groups of Z_{12}.

Solution: Subgroups of Z_{12} are $(\bar{0}), (\bar{2}), (\bar{3}), (\bar{4}), (\bar{6})$, and they are all cyclic with orders, respectively, 1, 6, 4, 3 and 2.

The quotient groups will also be cyclic of orders 12, 2, 3, 4 and 6, *i.e.* they will be isomorphic, respectively, to Z_{12}, Z_2, Z_3, Z_4 and Z_6.

3. Let K be an normal subgroup of G and $a \in G$. Show that the order of $aK \in G/K$ divides $o(a)$. Give an example to show that $o(aK)$ is a proper divisor of $o(a)$.

Solution: Let $o(a) = m$ so that $a^m = e$.

Now $(aK)^m = a^m K = eK = K$ so that $o(aK) \mid m$

Consider $G = S_3$ and $K = A_3$ and $\phi = \begin{pmatrix} 1 & 2 & 3 \\ 2 & 3 & 1 \end{pmatrix}$.

Then $o(\phi) = 3$, but $o(\phi K) = 1$ as $\phi K = K$

4. Show that the group $I(G)$ of inner automorphism of G is isomorphic to $G/Z(G)$. Deduce that, when G has trivial centre, G is isomorphic to a subgroup of Aut (G).

Solution: Consider the map $\theta : G \to I(G)$
$$a \to f_a$$

where $f_a(x) = axa^{-1}$. θ is clearly a homomorphism "onto" and Ker $\theta = Z(G)$.

Thus $$G/Z(G) \approx I(G). \text{ If } Z(G) = \{e\}$$
$$G \approx I(G) \subset \text{Aut}(G).$$

5. Show that if G has a trivial centre, Aut (G) also has trivial centre.

Solution: Since $Z(G) = \{e\}, G \approx I(G) \subset \text{Aut}(G)$

Let $\sigma \in Z(\text{Aut}(G))$. Then $\sigma f_a = f_a \sigma, \forall a \in G$

In particular $\sigma f_a(x) = f_a \sigma(x)$, for all $x \in G$

i.e. $\sigma(axa^{-1}) = a\sigma(x)a^{-1}$

and so $\sigma(a)\sigma(x)\sigma(a)^{-1} = a\sigma(x)a^{-1}$

This implies $a^{-1}\sigma(a)$ commutes with $\sigma(x)$ and since x is arbitrary and $\sigma \in \text{Aut}(G)$, $a^{-1}\sigma(a) \in Z(G) = \{e\}$. Then $a^{-1}\sigma(a) = e$ and so $\sigma(a) = a$. This implies $\sigma = Id$.

6. Let H and K be normal subgroups of G such that $H \cap K = \{e\}$. Show that $hk = kh$ for all $h \in H, k \in K$.

Solution: Consider $x = hkh^{-1}k^{-1}$

$$x = (hkh^{-1})k^{-1} \in K \text{ as } K \text{ is normal in } G$$
$$= h(kh^{-1}k^{-1}) \in H \text{ as } H \text{ is normal in } G$$

Since $H \cap K = \{e\}, x = e$ and so $hk = kh$.

7. Let H and K be subgroups of a finite group such that K is normal in G, and $G = HK$. Show that if $G/K \approx H$ then $H \cap K = \{e\}$.

Solution: Consider the map $\theta : H \to \dfrac{HK}{K} = \dfrac{G}{K}$ given by $\theta(h) = hK$. θ is a group homomorphism. If Ker $\theta \neq \{e\}$, G/K will be isomorphic to a quotient of H but it is given that G/K is isomorphic to H. Hence Ker $(\theta) = \{e\}$ and so $H \cap K = \{e\}$.

8. Let $\theta : G \to G'$ be a homomorphism of G in G'. If H is normal in G and H' normal in G' such that $\theta(H) \subset H'$, show that θ induces a homomorphism $\bar{\theta} : G/H \to G'/H'$. What is the Ker $\bar{\theta}$?

Solution: Consider the map $\bar{\theta} : G/H \to G'/H'$ defined by $\bar{\theta}(aH) = \theta(a)H'$. $\bar{\theta}$ is well defined as $\theta(H) \subset H'$ and is a homomorphism as θ is a homomorphism.

$$\text{Ker } \bar{\theta} = \{aH \mid \theta(a) \in H'\}$$
$$= \{aH \mid a \in \theta^{-1}(H')\}$$

9. Let $\theta : G \to G'$ be a homomorphism with $K = \text{Ker } \theta$. Show that θ induces a homomorphism $\bar{\theta} : G/K \to G'$ such that $\bar{\theta}\, p = \theta$, where $p : G \to G/K$ is the projection.

Solution: Define $\bar{\theta} : G/K \to G'$ by $\bar{\theta}(aK) = \theta(a)$. $\bar{\theta}$ is well defined as $aK = bK$ implies $b^{-1}a \in K = \text{Ker } \theta$ and $\theta(b^{-1}a) = e$, *i.e.* $\theta(b) = \theta(a)$. Moreover, $\bar{\theta}$ is a homomorphism as θ is a homomorphism and $\bar{\theta}\, p = \theta$.

10. Let m and n be integers such that $m \mid n$. Show that $\frac{Z_n}{\bar{m}\, Z_n} \approx Z_m$ where $\bar{m}\, Z_n = (\bar{m})$ is the cyclic group generated by $\bar{m}$ in Z_n.

Solution: Consider the map $\theta : Z_n \to Z_m$ defined by θ (a mod n) = a (mod m).

θ is well defined as $\bar{a} = \bar{b}$ implies $n \mid b - a$ and so $m \mid b - a$, *i.e.* a (mod m) = (b mod m).

Moreover θ is a homomorphism "onto" and Ker $\theta = (\bar{m})$. So $\frac{Z_n}{(\bar{m})} \approx Z_m$.

11. Let G be a group of order 18, H and K subgroups of G of orders 2 and 9. Show that $G = HK$.

Solution: K is normal in G as $(G : K) = 2$.

So HK is a subgroup of G and $H \cap K = \{e\}$ as its order must divide 2 and 9.

Thus $o(HK) = o(H)\, o(K) = 18$ and so $HK = G$.

12. If $G/Z(G)$ is cyclic, show that G is abelian.

Solution: Let $G/Z(G) = (\bar{x})$. For any $a, b \in G$ write $\bar{a} = \bar{x}^m$ and $\bar{b} = \bar{x}^n$.

Then $\quad a = x^m z_1$ and $b = x^n z_2$, $z_1, z_2 \in Z(G)$.

Then $\quad ab = x^m z_1 x^n z_2 = x^{m+n} z_1 z_2$

and $\quad ba = x^n z_2 x^m z_1 = x^{m+n} z_1 z_2$

So $\quad ab = ba$ and G is abelian.

1.9 DIRECT PRODUCTS

1. Let $G = H \times K$. Show that G is abelian if, and only if, H and K are abelian.

Solution: If G is abelian $(h_1, e)(h_2, e) = (h_2, e)(h_1, e)$ *i.e.* $(h_1 h_2, e) = (h_2 h_1, e)$. Thus $h_1 h_2 = h_2 h_1$ and H is abelian. Similarly K is abelian. Conversely if H and K are abelian, G is abelian because

$$(h, k)(h', k') = (hh', kk') = (h' h, k' k)$$
$$= (h', k')(h, k).$$

2. Let $G = H \times K$, where G is a finite group. Show that $o(G) = o(H)\, o(K)$.

Solution: $o(G)$ = cardinality of $\{(h, k) \mid h \in H, k \in K\}$

$$= o(H) \cdot o(K).$$

3. Let p and q be distinct primes. Show that Z_{pq} can be expressed as a direct product of a cyclic group of order p and a cyclic group of order q.

Solution: Consider the subgroups generated by $\bar{p}$ and $\bar{q}$ in Z_{pq} and let $K = (\bar{p})$ and $H = (\bar{q})$. Since $o(H) = p$ and $o(K) = q$, H and K are cyclic subgroups of orders p and q. Since p and q and distinct primes $H \cap K = \{e\}$. Also $G = H + K$ because there exist m, n such that $mp + nq = 1$.

For any $x \in Z$, $x = x\,mp + x\,nq$ and $\bar{x} = xm\bar{p} + xn\bar{q}$.

Hence $G = H + K$.

4. Investigate whether the following groups are cyclic.

(*i*) $Z_2 \times Z_2$. (*ii*) $Z_2 \times Z_3$.

Solution: (*i*) No as there is no element of order 4.

(*ii*) Yes as $(\bar{1}, \bar{1})$ is an element of order 6.

5. Express Z_{60} as a direct product of two subgroups.

Solution: Let $H = (\bar{5})$ and $K = (\overline{12})$. From the argument in Exercise 3 it follows that $Z_{60} = H \times K$.

6. Give an example of an infinite abelian group which cannot be written as a direct product of two proper subgroups.

Solution: Let $G = (Q, +)$ be rationals under addition. Suppose $G = H \oplus K$, H, K proper subgroups.

There exist $x \in H, x \neq 0$ and $y \in K, y \neq 0$.

Write $x = p/q$ and $y = s/t$. Then $qsx = pty \neq 0$ and belongs to $H \cap K$ a contradiction as $H \cap K = \{0\}$.

7. Show that the dihedral group D_4 cannot be expressed as a direct product of two proper subgroups (It is *indecomposable*).

Solution: $D_4 = \{e, a, a^2, a^3, b\; bc, ba^2, ba^3\}$.

If D_4 is expressed as a direct product of two proper subgroups, one of them must be of order 2 and the other of order 4. The only subgroup of order 4 is $H = (a)$. Let K be the other subgroup of order 2. Then K is abelian. If $D_4 = H \times K$, elements of H and K commute in a direct product and hence D_4 is abelian, which is a contradiction.

8. Let G be a direct product of two subgroups, each of which is cyclic of order 3. Show that G cannot be cyclic.

Solution: Let $G = H \times K$, $H \approx Z_3 \approx K$. Now $3\,(h, k) = (0, 0)$ and hence each non-trivial element of G will be of order 3, G cannot have any element of order 9.

9. Find the largest order of a cyclic group contained in $Z_6 \times Z_8$.

Solution: The subgroup $(\bar{2}) \times Z_8$ is cyclic as their orders 3 and 8 are relatively prime. It is the cyclic subgroup of largest order 24.

10. Let H and K be cyclic groups of relatively prime orders. Show that $G = H \times K$ is cyclic.

Solution: Let $o(H) = m$ and $o(K) = n$ with $(m, n) = 1$. Then $G = H \times K$ is cyclic with generator $(\bar{1}, \bar{1})$, because $mn\;(\bar{1}, \bar{1}) = (mn\,\bar{1}, mn\,\bar{1}) = (\bar{0}, \bar{0})$. Also if $d(\bar{1}, \bar{1}) = (\bar{0}, \bar{0})$, then $(\bar{d}, \bar{d}) = (\bar{0}, \bar{0})$. Thus $m \mid d$ and $n \mid d$. Since $(m, n) = 1$, $mn \mid d$ and $mn = d$. Thus G has an element of order mn and G is cyclic.

11. Let H and K be cyclic groups such that $G = H \times K$ is cyclic. Is it true that H and K have relatively prime orders?

Solution: Yes. $H \approx Z_m$ and $K \approx Z_n$ and $G \approx Z_m \times Z_n$. Then $o(G) = mn$.

Let $l = [m, n]$ be the l.c.m. and $d = [m, n]$ the g.c.d. of m and n. Then $ld = mn$.

Since G is cyclic G has an element $(\overline{a}, \overline{b})$ of order mn. Now $l\ (\overline{a}, \overline{b}) = (\overline{0}, \overline{0})$ as l is the l.c.m. of m and n. Hence $mn \mid l$. Clearly $l \mid mn$ and $l = mn$. Thus $d = 1$.

12. If p is a prime, show that the cyclic group Z_p^2 of order p^2 is indecomposable.

Solution: The only subgroups of Z_p^2 are 0, Z_p^2 and a subgroup H isomorphic to Z_p. Now Z_p^2 cannot be isomorphic to $H \times H$ as every non-trivial element of this group has order p. Thus Z_p^2 is indecomposable.

13. Show that the multiplicative group of non-zero complex numbers is the direct product of R^+, the multiplicative group of positive reals and R/Z, the quotient of the additive group of reals R modulo Z.

Solution: Consider the map $f : C^* \to R^+ \times R/Z$

$$re^{2\pi i\theta} \to (r, \overline{\theta})$$

f is easily verified to be a group isomorphism.

14. If N_i is normal in G_i $(i = 1, 2)$, show that $N_1 \times N_2$ is normal in $G_1 \times G_2$, and

$$\frac{G_1 \times G_2}{N_1 \times N_2} \approx \frac{G_1}{N_1} \times \frac{G_2}{N_2}.$$

Solution: If $(a, b) \in N_1 \times N_2$ and $(x, y) \in G_1 \times G_2$

then $(x, y)\ (a, b)\ (x, y)^{-1} = (xax^{-1}, yby^{-1}) \in N_1 \times N_2$

Thus $N_1 \times N_2$ is normal in $G_1 \times G_2$. The map $\theta : G_1 \times G_2 \to G_1/N_1 \times G_2/N_2$ defined by $\theta\ (x, y) = (\overline{x}, \overline{y})$ is a homomorphism "onto" with kernel $N_1 \times N_2$.

Hence $G_1 \times G_2/N_1 \times N_2 \approx G_1/N_1 \times G_2/N_2$.

1.10 PERMUTATION GROUPS

1. Which of the following permutations are (*i*) even (*ii*) odd?

(*i*) (1 2 3) (4 6 5), (*ii*) (12) (25 36) (13 24).

Solution: (*i*) Even (*ii*) Odd.

2. Express the inverse of the cycle (1 2 4 5 3) as a product of transpositions.

Solution: $(1\ 2\ 4\ 5\ 3)^{-1} = (3\ 5\ 4\ 2\ 1) = (31)\ (32)\ (34)\ (35)$.

3. Express the following permutations as products of disjoint cycles.

(*i*) $\begin{pmatrix} 1 & 2 & 3 & 4 & 5 & 6 & 7 & 8 \\ 2 & 3 & 4 & 5 & 7 & 6 & 8 & 1 \end{pmatrix}$ (*ii*) $\begin{pmatrix} 1 & 2 & 3 & 4 & 5 & 6 \\ 2 & 3 & 4 & 6 & 5 & 1 \end{pmatrix}$

Solution: (*i*) $\begin{pmatrix} 1 & 2 & 3 & 4 & 5 & 6 & 7 & 8 \\ 2 & 3 & 4 & 5 & 7 & 6 & 8 & 1 \end{pmatrix} = (1\ 2\ 3\ 4\ 5\ 7\ 8)$

(*ii*) $\begin{pmatrix} 1 & 2 & 3 & 4 & 5 & 6 \\ 2 & 3 & 4 & 6 & 5 & 1 \end{pmatrix} = (1\ 2\ 3\ 4\ 6)$

4. Show that an r-cycle is an element of order r in S_n.

Solution: Assume without loss of generality that $\sigma = (1\ 2\ 3\ \ldots\ r)$

Then $\sigma^2 = \begin{pmatrix} 1 & 2 & \ldots & r-1 & r \\ 3 & 4 & & 1 & 2 \end{pmatrix}$ and continuing

$$\sigma^{r-1} = \begin{pmatrix} 1 & 2 & \ldots & r \\ r & r-1 & & 1 \end{pmatrix} \neq Id$$

and $\sigma^r = Id$. So $o(\sigma) = r$.

5. Show that the converse of Lagrange's theorem is not true, by showing that A_4 has no subgroup of order 6.

Solution: $o(A_4) = \frac{1}{2}\underline{|4} = 12$

Let H be a subgroup of order 6 in A_4. Then H being of index 2 it is normal in A_4. Let $\sigma \in H$, $\sigma \neq Id$. Then σ can be expressed as a product of disjoint cycles. σ cannot be a cycle for then it has to be 3-cycle as σ is even and any normal subgroup H of A_4 containing a 3-cycle must be A_4. σ cannot be a product of 3 or more cycles as $n = 4$. σ can be a product of two transpositions. However, there are utmost 3 elements of this type. So H must contain a 3 cycle and then $H = A_4$, a contradiction.

6. Show that the smallest subgroup V of A_4 containing (1 2) (3 4), (1 3) (2 4) and (1 4) (2 3) is isomorphic to the Klein-4-group. Show further that it is a normal subgroup of A_4.

Solution: Let G be the Klein-4-group $G = \{e, a, b, c\}$ with

$$a^2 = b^2 = c^2 = e,\ ab = ba = c,\ bc = cb = a \text{ and } ca = ac = b.$$

Define a map $\theta : G \to V$ given by

$$\theta(a) = (1\ 2)(3\ 4),\ \theta(b) = (1\ 3)(2\ 4) \quad\text{and}\quad \theta(c) = (1\ 4)(2\ 3).$$

θ can be verified to be an isomorphism. Moreover V can be verified to be normal as $\sigma(1\ 2)(3\ 4)\sigma^{-1} = (\sigma(1)\ \sigma(2))(\sigma(3)\ \sigma(4))$ and must be an element of V. The same holds for the other elements.

7. Let $G = D_4$, $H = \{e, a^2, ab, a^3 b\}$ and $K = \{e, a^3 b\}$. Show that K is normal in H and H is normal in G but K is not normal in G.

Solution: Clearly H and K are subgroups. Since $(H : K) = 2$ and $(G : H) = 2$, K is normal in H and H is normal in G. But K is not normal in G because $b(a^3 b) b^{-1} = ba^3 = a^{-3}b = ab$ and this is not an element of K.

8. Show that every $\sigma \in A_n$ can be expressed as a product of 3 cycles.

Solution: σ can be expressed as a product of even number of transpositions. We can group these transpositions two by two. If they have a symbol in common their product is a 3 cycle as $(a_1, a_2)(a_1, a_3) = (a_1, a_3, a_2)$. If they do not have a symbol in common it is a product of two 3 cycles as

$$(a_1\ a_2)(a_3\ a_4) = (a_1\ a_2\ a_3)(a_2\ a_3\ a_4).$$

Thus each element is a product of 3 cycles.

9. Show that any normal subgroup H of A_n, containing a 3 cycle must coincide with A_n $(n \geq 5)$.

Solution: Let H be a normal subgroup of A_n containing a 3-cycle (a_1, a_2, a_3). Let (i, j, k) be any other 3 cycle and choose $\sigma \in S_n$ with

$$\sigma(a_1) = i, \sigma(a_2) = j, \sigma(a_3) = k.$$

Then $\quad \sigma(a_1, a_2, a_3)\sigma^{-1} = (i, j, k)$.

If σ is even this belongs to H. If σ is odd, choose two other symbols $a_4, a_5 \neq a_1, a_2, a_3$ $(n \geq 5)$ and let $\tau = (a_4, a_5)$ and replace σ by $\sigma\tau$.

Then $\quad \sigma\tau(a_1\ a_2\ a_3)(\sigma\tau)^{-1} = \sigma(a_1\ a_2\ a_3)\sigma^{-1} = (i, j, k)$.

Now $\sigma\tau$ is even and so $(i, j, k) \in H$. Hence the proof.

10. Show that any subgroup H of S_n, containing $\sigma = (12)$ and $\tau = (1\ 2\ 3 \ldots n)$ is S_n by showing $\tau^i\sigma\tau^{-i} = (i + 1, i + 2)$, $0 \leq i \leq n - 2$.

Solution: By computation $\tau^i\sigma\tau^{-i} = (i + 1, i + 2)$ $0 \leq i \leq n - 2$. So the subgroup, H contains all transpositions $(1\ 2)(2\ 3)(3\ 4), \ldots (n - 1, n)$. Hence it contains all transpositions $(1\ 3)(1\ 4) \ldots. (1 n - 1)(1\ n)$ because

$$(1\ 3) = (2\ 3)(1\ 2)(2\ 3)$$

$$(1\ 4) = (3\ 4)(1\ 3)(3\ 4), \text{ and so on.}$$

Thus H contains $(1\ 2)(1\ 3) \ldots (1\ n)$. But S_n is generated by these transposition as can be seen by induction on n. Assume that S_{n-1} is generated by $(1\ 2)(1\ 3) \ldots (1\ n - 1)$ and let $\sigma \in S_n$. If $\sigma(n) = m$, $(1n)(1m)\sigma$ leaves n fixed and so lies in S_{n-1}. Hence it is a product of transpositions $(1\ 2), \ldots (1n - 1)$. This implies σ is itself a product of $(1\ 2), \ldots. (1n)$. Thus $H = S_n$.

11. Show that S_n $(n \geq 3)$ has trivial centre.

Solution: Let $\sigma \in S_n$, $\sigma \neq e$. Then there exists some i, j, $i \neq j$ such that $\sigma(i) = j$. Since $n \geq 3$, choose $k \neq i, j$. Choose $\tau \in S_n$ such that $\tau(i) = i$ and $\tau(j) = k$.

Then $\quad \tau\sigma\tau^{-1}(i) = \tau\sigma(i) = \tau(j) = k.$

So $\quad \tau\sigma\tau^{-1} \neq \sigma \quad$ *i.e.* $\tau\sigma \neq \sigma\tau$.

Thus σ cannot be in the centre of S_n. Since σ is arbitrary, S_n has trivial centre.

12. Prove the general version on Cayley's theorem. If G is any group, then G is isomorphic to a subgroup of $A(G)$, where $A(G)$ is the group of all (1, 1) transformations of G onto G, with composition as group operation.

Solution: Define a map $\theta : G \rightarrow A(G)$

$$a \rightarrow f_a$$

where $f_a(x) = ax$, $x \in G$. Clearly θ_a is a group homomorphism and Ker $\theta = \{e\}$. Thus G is isomorphic is a subgroup of $A(G)$.

1.11 CONJUGATE CLASSES

1. Find all the conjugate classes of S_3 and verify the class equation.

Solution: $G = S_3 = \{e, \phi, \phi^2, \psi, \phi\psi, \psi\phi\}$.

The conjugate classes are $\{e\}$, $\{\phi, \phi^2, \phi\psi\}$ and $\{\psi, \psi\phi\}$ and have the number of elements 1, 3 and 2 which add up to 6, the order of G.

2. If G is a finite group, such that every element of G is of order of a certain power of a prime p, show that $o(G) = p^n$, for some $n \geq 1$.

Solution: Suppose $o(G)$ has a prime factor $q \neq p$. Then, by Cauchy's theorem G has an element of order q, a contradiction.

3. Give an example of an infinite group in which every element is of order of a certain power of a fixed prime p.

Solution: Let $G = \bigcup_{n=0}^{\infty} Z_{p^n}$, where Z_{p^n} is the group of p^nth roots of unity.

Since $Z_{p^n} \subset Z_{p^{n+1}}$, G is a group and has the required property.

4. Find all groups of order 9.

Solution: Since $o(G) = 9 = 3^2$, G is abelian.

Thus $G \approx Z_9$ or $Z_3 \times Z_3$.

5. Show that any non-abelian group of order 6 is isomorphic to S_3.

Solution: G is non-abelian. By Cauchy's theorem G has an element of order 3 say ϕ and an element of order 2 say ψ. Then $\phi\,\psi \neq \psi\,\phi$, otherwise $o(\phi\psi) = 6$ which is impossible. We can list the distinct elements of G as

$$G = \{e, \phi, \phi^2, \psi, \psi\phi, \phi\psi\}.$$

Now $\psi\phi^2 \in G$ and cannot be equal to any of the first 5 elements. So $\psi\phi^2 = \phi\psi$, *i.e.*, $\phi\psi = \psi\phi^{-1}$. Thus G is isomorphic to S_3.

6. If G is a non-abelian group of order p^3, show that $Z(G)$ is cyclic of order p.

Solution: Now $Z(G) \neq \{e\}$ and $Z(G) \neq G$ as G is non-abelian. So either $o(Z(G)) = p$ or p^2.

Now $o(Z(G)) = p^2$ is impossible for in that case, choose $a \in G$, $a \notin Z(G)$ and consider $N(a)$ the normaliser of a. Then $N(a) \supset Z(G)$ properly as $a \in N(a)$ and $a \notin Z(G)$. So $N(a) = G$, *i.e.* $a \in Z(G)$ a contradiction. So $o(Z(G)) = p$ and $Z(G)$ is cyclic of order p.

7. If G is a finite group with just 2 conjugate classes, show that $o(G) = 2$.

Solution: G has only one non-trivial conjugate class say C.

Let $o(G) = n$ so that by class equation, C has $n - 1$ elements. But $|C|$ divides n so that $n - 1 \mid n$, *i.e.* $n = 2$.

8. Let G be a group of order p^n. Show that G has a subgroup of order p^m, $0 \leq m \leq n$.

Solution: Induction on n, $n = 1$ is trivial. Assume the result for $n - 1$ and consider $Z(G)$. $Z(G) \neq \{e\}$ and choose an element $a \in Z(G)$ of order p. Then

$K = (a)$ is a normal subgroup of G and $G' = G/K$ has order p^{n-1}. By induction it has subgroups of the type H/K of orders p^m, $0 \le m \le n-1$. Then H is a subgroup of G of order p^{m+1}, $0 \le m \le n-1$ and $\{e\}$ has order $p^\circ = 1$.

9. Show that any subgroup of order p^{n-1} in a group G of order p^n, is normal in G.

Solution: Let H be a subgroup of order p^{n-1} in G.

Since $Z(a) \neq \{e\}$, choose $a \in Z(G)$, $a \neq e$ and let $K = (a)$.

Case I: If $a \in H$, $K \subset H$ and induction applies to G/K.

Then H/K is a group of order p^{n-2} in G/K and hence normal in G/K. So H is normal in G.

Case II: $a \notin H$. Consider $H \subset N(H) \subset G$.

Since $a \in Z(G) \subset N(H)$ and $a \notin H$, $H \not\subset N(H)$.

So $N(H) = G$ and hence H is normal in G.

10. Let G be a group of order p^n, and H a non-trivial normal subgroup. Show that $H \cap Z(G) \neq \{e\}$.

Solution: Since H is normal in G, if a conjugate class of G intersects H, it must be contained in H. So H is a union of those conjugate classes which intersect H. Some of these will be trivial classes. Writing class equation for H we have

$$o(H) = H \cap Z(G) + \sum_{|G:N(a)|>1} |G : N(a)|$$

where the summation is over those non-trivial conjugate classes intersecting H. p divides the sum and $p \mid o(H)$.

Hence $p \mid o(H \cap Z(G))$.

In particular, $H \cap Z(G) \neq \{e\}$.

11. Show that any abelian group of order 21 is cyclic.

Solution: Let $o(G) = 21$. By Cauchy's theorem G has elements a, b with $o(a) = 3$ and $o(b) = 7$. Since $ab = ba$, $o(ab) = 21$. Hence G is cyclic.

12. Let G be a cyclic group of order n such that for every pair of distinct sub groups H, K of G either $H \subset K$ or $K \subset H$. Show that n is a prime power.

Solution: Let $o(G) = p^\lambda t$, $(t, p) = 1$. If possible let t have a prime factor q. By Cauchy's theorem G has a subgroup H of order p and a subgroup K of order q. Since $p \neq q$, $H \subset K$ or $K \subset H$ is not possible. So $t = 1$ and $o(G) = p^\lambda$.

13. Let G be any group. Show that the set H of all elements of G which have only finitely many conjugates is a subgroup of G. Is this subgroup normal?

Solution: Let $a, b, \in G$ such that the number of conjugates of a and b are m and n respectively. Then $xabx^{-1} = (xbx^{-1})(xbx^{-1})$. So the conjugates of ab are utmost mn.

Also $x^{-1}a^{-1}x = (x^{-1}ax)^{-1}$ and so there are m conjugates of a^{-1}. Hence, the set H is a subgroup. This subgroup will be normal because it will be left invariant by any automorphism of G. If $a \in H$ and $\sigma \in \text{Aut}(G)$, then $\sigma(a)$ will have only finitely many conjugates as

$$x^{-1}\sigma(a)\,x = \sigma(y)^{-1}\,\sigma(a)\,\sigma(y) \text{ (where } \sigma(y) = x)$$
$$= \sigma(y^{-1}\,a\,y),\ \sigma \text{ times a conjugate of } a.$$

In particular H will be invariant under inner automorphisms, *i.e.* H is normal.

14. Show that a group of order $2m$, m odd has a subgroup of index 2.

Solution: We first prove the following:

It G is a subgroup of S_n which contains an odd permutation, then G contains a subgroup of index 2. Now A_n is a normal subgroup of S_n with $(S_n : A_n) = 2$ and since $G \not\subset A_n$, $GA_n = S_n$.

Hence $\quad (GA_n : A_n) = 2.$

But $\quad (GA_n : A_n) = (G : G \cap A_n) = 2.$

Hence $G \cap A_n$ is normal subgroup of index 2 in G.

Now consider the given group G of order $2m = n$, m odd. Consider the regular representation of $\theta : G \to S_n$ given by $\theta(a) = \rho_a$, where $\rho_a(x) = ax$. This is an isomorphism of G into S_n. By Cauchy's theorem G has an element a of order 2 and let $\theta(a) = \sigma \cdot \sigma \in S_n$ is of order 2 and hence σ is a product of transpositions. Since σ does not fix any symbol, and $n = 2m$, σ is a product of m transpositions. Since m is odd, σ is odd. By the first part $\theta(G) = G' \subset S_n$ contains a subgroup of index 2 as $\sigma \in G'$. Its inverse image under θ is the required subgroup of index 2 in G.

15. Let $\sigma = (1, 2, 3, ..., n) \in S_n$. Find the number of elements in the conjugate class of σ.

Solution: Any conjugate of σ is of the type

$$\tau\sigma\tau^{-1} = (\tau(1), ..., \tau(n))$$

and it is also an n-cycle. The number of conjugates of σ is equal to the number of distinct n cycles. Any permutation on n symbols defines an n-cycle with cyclic shifts defining the same n-cycles.

Hence, the number of distinct conjugates is equal to $\dfrac{\underline{|n}}{n} = \underline{|n-1}$.

1.12 SYLOW THEOREMS AND SOLVABLE GROUPS

1. Let G be a finite cyclic group of order n. Show that for any divisor d of n, G has a unique subgroup of order d.

Solution: Let $G = (a)$, $o(a) = n$ and let $b = a^{n/d}$. Then $H = (b)$ is cyclic of order d.

H is unique for let H' be another subgroup of order d, $H' = (a^i)$.

Then $o(a^i) = d$ and so $i = n/d$.

2. Show that a group of order 108 and a group of order 148 cannot be simple.

Solution: If G is a group, H a subgroup of G and $S = G/H$, the set of right cosets of H in G, the map $\theta : G \to A(S)$ defined by $\theta(a) = \rho_a$, left multiplication by a is a group homomorphism with kernel, largest normal subgroup contained in H. In particular, θ cannot be an isomorphism if $o(G) \not| \; |i(H)|$.

Case I: Let $o(G) = 108 = 2^2 \cdot 3^3 \cdot G$ contains a 3-sylow subgroup H and index of H in G is $2^2 = 4$. Since $o(G)$ does not divide $|i(H)|$, H contains a non-trivial normal subgroup. Hence G is not simple.

Case II: Let $o(G) = 148 = 2^2 \cdot 37 \cdot G$ has a sylow 37 subgroup and the number of such subgroups is $1 + 37t$ and it divides the index 4.

So $1 + 37t \mid 4$ implies $t = 0$. G has a unique sylow 37 group which must be normal. So G is not simple.

3. Let G be a group of order $2p$, p prime. Show that either G is cyclic or G is generated by $\{a, b\}$ with relations $a^p = e = b^2$ and $bab = a^{-1}$.

Solution: Since $o(G) = 2p$, G has a sylow p-subgroup H and the number of such subgroups is $1 + tp \mid 2$. So $t = 0$, *i.e.* G has a unique p-sylow subgroup H which is cyclic say $H = (a)$. Now G/H is cyclic of order 2. Choose $b \in G$, $b \notin H$. Then G has elements $\{e, a, a^2, ... a^{p-1}, b, ba, ... ba^{p-1}\}$. If $o(b) = i$. Then $i \mid 2p$ and so $i = 2$ or $i = p$. $i = p$ is impossible as then b will lie in H, the unique subgroup of order p. So $o(b) = 2$ *i.e.* $b^2 = e$. Now consider bab^{-1}. It belongs to H as H is normal and so $bab^{-1} = a^i$.

So $\qquad a = ba^i b^{-1} = (bab^{-1})^i = (a^i)^i = a^{i^2}$.

So $\qquad i^2 = 1 \pmod p \quad$ *i.e.* $\quad p \mid (i+1)(i-1)$

So $\qquad i + 1 = p \quad$ or $\quad i - 1 = p \quad$ i.e. $\quad i = p - 1 \quad$ or $\quad p + 1$.

If $\; i = p - 1, \qquad bab^{-1} = a^i = a^{-1}$

If $\; i = p + 1, \qquad bab^{-1} = a^{p+1} = a \quad$ *i.e.* $\quad ab = ba$.

In the first case G is non-abelian generated by a, b with relations $a^p = e = b^2$ and $bab^{-1} = a^{-1}$ *i.e.* $bab = a^{-1}$.

In the second case G is abelian and $o(ab) = 2p$, *i.e.* G is cyclic of order $2p$.

4. Prove that any solvable group G has always an abelian normal subgroup K, $K \neq \{e\}$.

Solution: It G is a solvable group, there exists n such that $G^{(n)} = \{e\}$. $n \geq 1$. If G is abelian we can take $K = G$. Assume G is non-abelian. We claim that each $G^{(i)}$ is normal in G, in fact it is a characteristic subgroup of G *i.e.* $\sigma(G^{(i)}) \subset G^{(i)}$ for every $\sigma \in \text{Aut}(G)$.

Clearly $\sigma(G^{(1)}) \subset G^{(1)}$ because

$$\sigma(aba^{-1}b^{-1}) = \sigma(a)\,\sigma(b)\,\sigma(a)^{-1}\,\sigma(b)^{-1}.$$

By iteration $\sigma(G^{(1)}) \subset G^{(1)}$. In particular, $G^{(i)}$ is left invariant by inner automorphisms of G and hence $G^{(i)}$ is normal in G. Now $G^{(n-1)}$ is normal in G and is abelian because $G^{(n)} = \{e\}$. So $K = G^{(n-1)}$ is the required group.

5. Let H and K be normal solvable subgroups of G. Show that HK is solvable.

Solution: Now HK is a subgroup of G and $HK/K \approx H/H \cap K$ is solvable as H is solvable.

Since HK/K and K are solvable. HK is solvable.

6. Let G be a group, σ a non-trivial automorphism of G such that for each $x \in G$, either $\sigma(x) = x$ or $\sigma(x) = x^{-1}$. Show that G is solvable.

Solution: Let $H = \{x \in G \mid \sigma(x) = x\}$. Since $\sigma \neq Id$, $H \neq G$. Let $N(H)$ be the normaliser of H in G. We claim that $G/N(H)$ is abelian, by showing that the square of every element is identity. Let $x^2 \in G$, $x^2 \notin H$.

Then for any $h \in H$, $\sigma(x^2 h) = x^2 h$ or $(x^2 h)^{-1}$.

If $\sigma(x^2 h) = x^2 h$, *i.e.* $\sigma(x^2) h = x^2 h$, $\sigma(x^2) = x^2$, then $x^2 \in H$ a contradiction.

Hence $\quad \sigma(x^2 h) = (x^2 h)^{-1} = h^{-1} x^{-2}$

i.e. $\quad \sigma(x^2) h = x^{-2} h = h^{-1} x^{-2}$ and so $hx^2 = x^2 h^{-1}$

i.e. $\quad Hx^2 = x^2 H$, $x^2 \in N(H)$.

Thus the square of every element of $G/N(H)$ is identity.

We now claim that $N(H)/H$ is abelian. The restriction $\bar{\sigma}$ of σ to $N(H)/H$ is an automorphism such that $\bar{\sigma}(\bar{x}) = \bar{x}^{-1}$. Hence $N(H)/H$ is abelian.

We now claim that H is abelian by showing that three is an inner automorphism of H which is the same as $h \to h^{-1}$.

Choose $y \in N(H)$, $y \notin H$. Then i_y is an automorphism of $N(H)$ as H is normal in $N(H)$.

If $h \in H$, then $yh \notin H$ and so $\sigma(yh) = (yh)^{-1}$. But $\sigma(y) = y^{-1}$ and $\sigma(h) = h$ so that $y^{-1} h = h^{-1} y^{-1}$ *i.e.* $y^{-1} hy = h^{-1}$. Thus i_y is the same as $h \to h^{-1}$. Hence H is abelian.

G has a normal series $G \supset N(H) \supset H \supset \{e\}$ with abelian quotients. Hence G is solvable.

7. True or false? Justify. If every proper subgroup of G is solvable then G is solvable.

Solution: False. $G = A_5$ is not solvable for if it is solvable S_5/A_5 being solvable, S_5 will be solvable which is not the case. We claim that all proper subgroup of A_5 are solvable. Now $o(A_5) = 60 = 2^2 \cdot 3 \cdot 5 \cdot A_5$ is simple and hence it has no subgroup of order 30, for then it will be normal in A_5. The subgroups of A_5 can have only orders 2, 2^2, 3, 5, 2.3, 2.5, 2^2.3, 2^2.5 and 3.5. All these groups are solvable for, if p, q are distinct primes, groups of order $p^2 q$ and pq are solvable. (Hard).

8. Find all sylow 3-subgroups of S_4 and verify that they are all conjugate.

Solution: $\quad o(S_4) = 24 = 2^3 . 3$.

The number of sylow 3-subgroups $1 + 3t \mid 8$. So $t = 0$ or $t = 1$ are the only possibilities. Let $t = 0$. The sylow 3-subgroup is unique and hence normal. Let $t = 1$, there are 4 sylow 3-subgroups and they are all cyclic of order 3. *i.e.* generated by the 3 cycles (1 2 3), (1 2 4), (1 3 4) and (2 4 3). Obviously these are all conjugate to each other and they are distinct.

9. Show that no group of order less than 60 and not a prime can be simple.

Solution: For any such group G, write the $o(G)$ as a product of prime factors, where there must be more than one prime factor as $o(G)$ is not a prime. In each case we can choose a sylow p-subgroup H for a suitable prime factor p, such that $\underline{|i(H)} < o(G)$. In particular $o(G) \chi \underline{|i(H)}$ and G cannot be simple.

10. Show that the group of all matrices of the type

$$\begin{pmatrix} 1 & a & b \\ 0 & 1 & c \\ 0 & 0 & 1 \end{pmatrix}, a, b, c \in Z_3$$

is a non-abelian group for multiplication and that it is solvable.

Solution: Multiplying two elements of the given type of matrix gives another element of the same type and the multiplication is not commutative. Moreover *Id* is of the given type, corresponding to $a = b = c = 0$ and we can write the inverse of the matrix which is also of the same type. Hence these matrices form non-abelian group. This group is solvable because a, b, c being arbiurary elements of Z_3, there are in all 27 choices for the values of these elements. Hence G is a group of order 27 which is the power of a prime. Hence G is solvable.

11. Show that a direct product of two nilpotent groups is nilpotent.

Solution: Let $G = G_1 \times G_2$, where G_1 and G_2 are nilpotent.

Then $$Z^{(1)} = [G, G] = [G_1 \times G_2, \times G_1 \times G_2]$$
$$= [G_1, G_1] \times [G_2, G_2] = Z_1^{(1)} \times Z_2^{(1)}$$

By iteration $$Z^{(i)} = Z_1^{(i)} \times Z_2^{(i)}$$

Since G_1 and G_2 are nilpotent.

$$Z_1^{(m)} = Z_2^{(m)} = \{e\} \text{ for some } m.$$

Hence $Z^{(m)} = \{e\}$. This implies that G is nilpotent.

1.13 JORDAN HOLDER THEOREM

1. Write down two composition series for Z_{30} and verify that they are equivalent.

Solution: $Z_{30} \supset (\bar{2}) \supset (\bar{6}) \supset \{0\}$ is a composition series of Z_{30} and $Z_{30} \supset (\bar{3}) \supset (\bar{6}) \supset \{0\}$ is another composition series. They are equivalent because the quotients are cyclic groups of orders 2, 3 and 5 and 3, 2 and 5 respectively.

2. Show that an abelian group has a composition series if, and only if, it is finite.

Solution: Let G be abelian and let $G \supset G_1 \supset ... \supset G_n = \{e\}$ be a composition series of G. Then each G_i/G_{i+1} is simple abelian and hence cyclic of prime order. If $o(G_{i+1}) < \infty$ then $o(G_i) < \infty$. Since $o(G_n) = 1$, it follow that $o(G) < \infty$. Conversely if G is finite abelian it has a composition series by induction on

$o(G) = n$. Let p be a prime dividing n and choose a subgroup G_1 of G such that $o(G_1) = n/p$. Then G_1 has a composition series $G_1 \supset G_2 \supset \ldots \supset G_n = \{e\}$ and then $G \supset G_1 \supset G_2 \supset \ldots \supset G_n = \{e\}$ is a composition series of G.

3. Give examples of groups G and G_1 such that they have composition series whose quotients are isomorphic in some order but G and G_1 themselves are not isomorphic.

Solution: Let $G = Z_6$ and $G_1 = S_3$.

Then $G \supset (\bar{2}) \supset \{0\}$ is a composition series

and $S_3 \supset A_3 \supset \{e\}$ is a composition series.

The quotients are isomorphic in some order but G is not isomorphic to S_3.

4. Find a composition series of $S_3 \times Z_2$.

Solution: $G = S_3 \times Z_2 \supset A_3 \times Z_2 \supset \{e\} \times Z_2 \supset \{e\} \times \{e\}$ is a composition series of G.

5. Show that $S_n \supset A_n \supset \{e\}$ is a normal series for $n > 4$. Deduce that S_n is not solvable for $n > 4$.

Solution: Since A_n is normal in S_n, the series is normal. If S_n is solvable, so is A_n. But A_n is not solvable for $n > 4$ as A_n is simple for $n > 4$ and a solvable group has always a non trivial normal subgroup.

6. Find a composition series of $S_3 \times S_3$. Is this group solvable? Why?

Solution: $S_3 \times S_3 \supset A_3 \times S_3 \supset A_3 \times A_3 \supset A_3 \times \{e\} \supset \{e\} \times \{e\}$ is a composition series with quotients cyclic of orders 2, 2, 3 and 3 all cyclic of prime order. The group is solvable as all quotients are abelian.

7. Let $G \supset G_1 \supset G_2 \supset \ldots \supset G_n = \{e\}$ be a composition series of G and K a simple normal subgroup of G. Show that

$$G \supset G_1 K \supset G_2 K \supset \ldots \supset K \supset \{e\}$$

is a composition series of G after omitting the repeated factors.

Solution: Since G_{i+1} is normal in G_i, $G_{i+1} K$ is normal in $G_i K$ and the quotient $G_i K/G_{i+1} K \approx G_i/G_{i+1}$ if $G_i K \neq G_{i+1} K$.

For this consider the projection maps

$$G_i \to G_i K/K \to \frac{G_i K/K}{G_{i+1}K/K}$$

which is a surjective group homomorphism with kernel G_{i+1} as $G_i K \neq G_{i+1} K$.

Hence $G_i/G_{i+1} \approx \dfrac{G_i K}{G_{i+1} K}$ after omitting repeated factors.

Hence $G_i K/G_{i+1} K$ is simple.

Also since K is simple, $K \{e\}$ is simple. Hence, the result.

CHAPTER 2

Rings

2.1 RINGS

1. Examine whether the following satisfy the axioms for a ring:

   (*i*) The set of positive integers for + and •

   (*ii*) The set of all rational numbers of the form m/n, m, n having no common factors, and p does not divide n (p being a fixed prime) for usual + and •

   (*iii*) The set of all vectors in R^3 with usual addition of vectors and • as cross product.

   Solution:

   (*i*) No. Three is no additive inverse

   (*ii*) Yes.

   (*iii*) No. Cross product is not associative.

2. Given an example of a ring which is neither commutative nor has a unit element.

   Solution: All $n \times n$ matrices with coefficients in the ring of even integers.

3. Show that $a \cdot (b - c) = a \cdot b - a \cdot c$, $a, b, c \in R$

   Solution: $a \cdot (b - c) = a \cdot (b + (-c)) = a \cdot b + a \cdot (-c)$

   $= a \cdot b - (a \cdot c)$

4. Given an example of a ring with finite number of elements which is not commutative.

   Solution: The set of all 2 × 2 matrices with entries in the field Z_2.

5. Let R be a ring, $a, b \in R$. Show that for any positive integer n, $(na)\, b = n(ab) = a(nb)$.

   Solution: $(na)\, b = (a + a + \ldots\ n \text{ times}) \cdot b$

   $= a \cdot b + a \cdot b + \ldots\ n \text{ times} = n\,(ab)$

   $a \cdot (nb) = a\,(b + b + \ldots\ n \text{ times})$

   $= a\ b + a \cdot b + \ldots\ n \text{ times} = n\,(ab)$

6. Let X be any set and R a ring. Show that $C(X, R)$ which is the set of all functions $f: X \to R$ is a ring for the operations, $(f+g)(x) = f(x) + g(x)$, $(fg)(x) = f(x) \cdot g(x)$, $f, g \in C(X, R)$, $x \in X$. Is the ring commutative? Does it have a unit element?

Solution: The axioms for the ring can be verified easily with zero element as the constant zero function. The ring is commutative if R is commutative and has a unit element if R has unit element, viz. the constant function 1.

7. Give an example of a ring in which $a^2 = a$ for all $a \in R$. (Such a ring is called a *Boolean ring*).

Solution: Let X be any set and $R = P(X)$ the set of all subsets of X. R is a ring for the operations.

$$A + B = A \Delta B = (A - B) \cup (B - A)$$

$$A \times B = A \cap B.$$

The zero element is the null set ϕ and the additive inverse of A is A. In this ring

$$A^2 = A \cap A = A.$$

8. Show that in a Boolean ring $2a = 0$, $a \in R$. Deduce that a Boolean ring is commutative.

Solution: $(a + b)^2 = a + b$

$$(a + b)^2 = (a + b)(a + b) = a^2 + ab + ba + b^2 = a + b$$

So $\quad ab + ba = 0 \quad$ for all $a, b \in R$

Put $\quad b = a;\ a^2 + a^2 = a + a = 2a = 0$, for all $a \in R$.

Hence $\quad (-a) = a$ for all a

$ba = -(ab) = ab$. Hence R is commutative.

9. Let $R = \{0, 1\}$ with + and $\cdot$ defined by

$$0 + 0 = 0 = 1 + 1, \quad 1 + 0 = 0 + 1 = 1$$

$$0 \cdot 0 = 0 = 1 \cdot 0 = 0 \cdot 1, \quad 1 \cdot 1 = 1.$$

Show that (*a*) R is a ring and (*b*) it is a Boolean ring.

Solution: The axioms for the ring can be verified easily with 0 as zero element. Clearly $a^2 = a$, for all $a \in R$. Hence R is a Boolean ring.

10. If R is a ring, show that $(a + b)^2 = a^2 + ab + ba + b^2$ and that when R is commutative $\quad (a + b)^2 = a^2 + 2ab + b^2$

Solution: $\quad (a + b)^2 = (a + b)(a + b) = a^2 + ab + ba + b^2$

If R is commutative, $ab = ba$ so that

$$(a + b)^2 = a^2 + 2ab + b^2.$$

11. Prove by induction on n, the binomial theorem in a commutative ring,

i.e. $\quad (a + b)^n = a^n + \binom{n}{1} a^{n-1} b + ... + b^n$, $a, b \in R$.

Solution: The proof is induction as n. For $n = 2$, the result is true by the above problem. Assume the result for $n - 1$.

$$(a + b)^{n-1} = a^{n-1} + \binom{n-1}{1} a^{n-2} b + ... + b^{n-1}$$

$$(a + b)^n = (a + b)(a + b)^{n-1}$$
$$= a(a + b)^{n-1} + b(a + b)^{n-1}$$
$$= a\left[a^{n-1} + \binom{n-1}{1} a^{n-2} b + ... + b^{n-1}\right]$$
$$+ b\left[a^{n-1} + \binom{n-1}{1} a^{n-2} b + ... + b^{n-1}\right]$$
$$= a^n + \left\{\binom{n-1}{1} + 1\right\} a^{n-1} b + \left\{\binom{n-1}{2} + \binom{n-1}{1}\right\} a^{n-2} b^2 + ... + b^n$$
$$= a^n + \binom{n}{1} a^{n-1} b + \binom{n}{2} a^{n-2} b^2 + ... + b^n$$

2.2 INTEGRAL DOMAINS

1. Which of the following rings are integral domains? Fields?

   (*i*) $R = \{a + b\sqrt{-5} \mid a, b \in Z\}$ for + and · as complex numbers.

   (*ii*) For a fixed prime p, $R_p = \{| a/b \mid a, b \in Z$, without common factor and p does not divide $b\}$ with respect to + and · in Q.

   Solution: (*i*) Yes, it is an integral domain because C is an integral domain.

   (*ii*) Yes it is an integral domain because Q is an integral domain.

2. Show that a finite ring with unity and without zero divisors is a skew field.

   Solution: Let R be a finite ring without zero divisors, and let the non-zero elements be $a_1, ..., a_n$. For any $a \neq 0$, $aa_1, ..., aa_n$ are distinct for if $aa_i = aa_j$, $a(a_i - a_j) = 0$, and hence $a_i - a_j = 0$, *i.e.* $a_i = a_j$. Since $1 \in R$ we solve $aa_k = 1$ for some k. We have shown that every non-zero element has a right inverse. Similarly we can show that every non-zero element has a left inverse. This will imply that both left and right inverses are the same and $a \neq 0$, will have a unique inverse. Hence R is a skew field.

3. For any ring R, define the centre $Z(R)$ as

$$Z(R) = \{a \mid a \in R, \quad ax = xa, \quad \text{for all } x \in R\}$$

   Show that $Z(R)$ is a ring for the induced operations of + and

   Solution: If $a, b \in Z(R)$, then $(a + b)x = ax + bx$

$$= xa + xb = x(a + b), x \in R.$$

Thus $a + b \in Z(R)$. Also $(ab)\,x = a\,(bx)$

$$= a\,(xb) = (ax)\,b = (xa)\,b = x\,(ab).$$

Hence $ab \in Z(R)$ and thus $Z(R)$ is a ring.

4. Show that if R is a skew field, $Z(R)$ is a field.

Solution: Let R be a skew field.

If $a \in Z(R)$ $ax = xa, x \in R$. Let $x \neq 0$.

$$x = 1 \cdot x = a^{-1}\,(ax) = a^{-1}\,(xa) = (a^{-1}\,x)\,a$$

Therefore $xa^{-1} = a^{-1}\,x$ and so $a^{-1} \in Z(R)$.

Thus $Z(R)$ is a skew field and since it is commutative, $Z(R)$ is a field.

5. An element $a \in R$ is called *nilpotent* if $a^n = 0$ for some $n \geq 1$. Give examples of non-zero nilpotent elements in (*i*) ring of 2×2 matrices, (*ii*) ring Z_8, (*iii*) show that an integral domain has no non-zero nilpotent elements.

Solution: (*i*) $A = \begin{pmatrix} 0 & 1 \\ 0 & 0 \end{pmatrix}$ satisfies $A^2 = \begin{pmatrix} 0 & 0 \\ 0 & 0 \end{pmatrix}$

(*ii*) $\overline{2} \in Z_8$ satisfies $(\overline{2})^3 = \overline{0}$

(*iii*) Let R be an integral domain, and $a \in R, a \neq 0$, such that $a^n = 0$. Choose the least such n.

Then $a^n = a \cdot (a^{n-1}) = 0.$

Since $a \neq 0$, $a^{n-1} = 0$, contradiction to be choice of n.

6. An element is called idempotent if $a^2 = a$. Give an example of non-zero idempotent elements in the ring 2×2 matrices. Give an example of a ring in which every element is idempotent. What are the idempotent elements in an integral domain with unit element?

Solution: (*i*) $A = \begin{pmatrix} 1 & 0 \\ 0 & 0 \end{pmatrix}$ satisfies $A^2 = A$.

(*ii*) In any Boolean ring, every element is idempotent. As for instance, in the ring Z_2.

(*iii*) Let R be an integral domain, $a \in R$ and satisfy $a^2 = a$. Then $a\,(a - 1) = a^2 - a = 0$. Then either $a = 0$, or $a = 1$. Thus 0 and 1 are the only idempotent elements in a domain.

7. Show that if a is a regular element in a ring, so is $(-a)$. Is the same true of the sum of two regular elements? Why? Show that if $1 - ab$ is a regular element, so is $1 - ba$.

Solution: Let $ab = 1 = ba$.

Then $(-a)\,(-b) = ab = 1 = (-b)\,(-a).$

Hence $-a$ is regular.

(*ii*) Sum of two regular elements is not regular for if a is regular, so is $-a$ and $a + (-a) = 0$ is not regular.

(*iii*) Since $1 - ab$ is regular, there exists c such that

$$c\,(1 - ab) = (1 - ab)\,c = 1.$$

We claim that the inverse of $1 - ba$ is $1 + bca$.

Consider $(1 - ba)(1 + bca)$

$$= 1 + bca - ba - babca$$
$$= 1 + bca - ba - b(c-1)a$$
$$= 1 + bca - ba - bca + ba$$
$$= 1$$

Similarly, $(1 + bca)(1 - ba)$

$$= 1 - ba + bca - bcaba$$
$$= 1 - ba + bca - b(c-1)a$$
$$= 1 - ba + bca - bca + ba$$
$$= 1$$

Hence $1 - ba$ is regular.

8. Let R be a commutative ring with unit element. Show that the regular elements form a group for multiplication (Group of units of R).

Solution: Let $U(R)$ be the set on regular elements of R. Clearly $1 \in U(R)$. If $a \in U(R)$, $ab = ba = 1$ for some $a \in R$, and $a^{-1} = b \in U(R)$. If $a, b \in U(R)$, then $ab \in U(R)$. Hence $U(R)$ is a group.

9. What are the group of units of (*i*) Z, (*ii*) Z_n and (*iii*) $\{a + b\sqrt{-5} \mid a, b \in Z\}$.

Solution: (*i*) $U(R) = \{\pm 1\}$

(*ii*) $U(R) = \{\bar{a} \mid a \in Z_n, (a, n) = 1\}$

(*iii*) $U(R) = \{\pm 1\}$.

10. Which elements of Z_p are their own inverses?

Solution: Let $\bar{a} \in Z_p$ such that $\bar{a}^2 = \bar{1}$. Then

$$a^2 \equiv 1 \pmod p \text{ i.e. } p \mid (a^2 - 1) = (a + 1)(a - 1).$$

$$p \mid a + 1 \text{ or } p \mid a - 1, \text{ i.e. } a \equiv 1 \pmod p$$

or $\quad a \equiv 1 \pmod p$. Hence $\bar{a} = \bar{1}$ or $\bar{a} = \overline{p-1}$.

11. What is the characteristic of the following rings?

(*i*) $3Z$ (*ii*) Z_3 (*iii*) $\{a + b\sqrt{-5} \mid a, b \in Z\}$.

Solution: (*i*) Zero (*ii*) Three (*iii*) Zero.

12. Show that for every $n \geq 0$, there is a ring with characteristic n.

Solution: $R = Z_n$

13. If ch $R = p$, p prime, show that $(a + b)^p = a^p + b^p$, in a commutative ring R, a, $b \in R$.

Solution:

$$(a + b)^p = a^p + \binom{p}{1} a^{b-1} b + \dots + \binom{p}{p-1} ab^{p-1} + b^p.$$

Since $p \left| \binom{p}{i}, i \geq 1, \binom{p}{i} \right. = 0$. Hence $(a + b)^p = a^p + b^p$.

14. Show that a finite field cannot be of characteristic zero.

Solution: Let ch $F = 0$. Consider the map $\theta : Z \to F$ such that $\theta\ (1) = 1_F$, unit element of F. θ is a ring homomorphism, which is injective as ch $F = 0$. In particular F is infinite.

15. Show that it is impossible to have an integral domain with 6 elements.

Solution: Consider the additive abelian group R. Clearly R has elements of order 2 and order 3, as since the group is abelian, the sum element has order 6. This implies, ch $R = 6$, a contradiction, since the characteristic of an integral domain is always a prime.

16. Show that any ring R with 6 elements is commutative.

Solution: Consider the additive group of R. It is an abelian group of order 6. Hence, it has elements of orders 2 and 3 and hence an element of order 6, as $(2, 3) = 1$ and the group is abelian. Thus the additive group of R is cyclic and is of the form $R = \{0, a, 2a, ..., 5a\}$. The product of any two element depends on the value of a^2 and hence the product is commutative, whatever be the value of a^2.

2.3 ISOMORPHISM

1. Show that the ring Z and $2Z$ are not isomorphic.

Solution: Z has a unit element but $2Z$ has no unit element.

2. Are the following mappings isomorphisms of rings? Why?

(*i*) R is the ring of real quaternions and $f : R \to R$ is given by $f(a + bi + cj + dk) = a - bi - cj - dk$.

(*ii*) R is the ring of 2×2 matrices and $f : R \to R$ given by

$$f\begin{pmatrix} a & b \\ c & d \end{pmatrix} = \begin{pmatrix} a & c \\ b & d \end{pmatrix}$$

(*iii*) $R = \{a + b\sqrt{-5} \mid a, b \in Z\}$ and

$$R' = \{a + b\sqrt{-2} \mid a, b \in Z\}$$

$f : R \to R'$ defined by $f\ (a + b\sqrt{-5}) = a + b\sqrt{-2}$.

Solution: (*i*) Yes.

(*ii*) No as f does not preserve multiplication.

(*iii*) No as f does not preserve multiplication.

3. Find all rings up to isomorphism with (*i*) two elements (*ii*) three elements.

Solution: (*i*) Let $R = \{0, a\}$, $a \neq 0$. Then $a + a = 0$ and either $a^2 = 0$, in which case R is a ring on additive group with trivial multiplication or $a^2 = a$, in which case R is isomorphic to Z_2.

(*ii*) $R = \{0, a, b\}$. Then $2a = b$ and $3a = 0$.

Thus $R = \{0, a, 2a\}$. If $a^2 = 0$, R is the ring on the additive group of 3 elements with trivial multiplication.

If $a^2 = a$, then a is the unit element of the ring and there are no zero divisors. Hence R is a field isomorphic to Z_3.

If $a^2 = 2a$, then $2a$ is the unit element of the ring and the ring has no zero divisors. Hence R is a field isomorphic to Z_3.

4. Consider the ring R and R' with four elements $R = \{a, b, c, d\}$ and $R' = \{x, y, z, t\}$ and $+, \cdot$ defined by the following tables. Investigate whether R and R' are isomorphic.

+	a	b	c	d
a	a	b	c	d
b	b	a	d	c
c	c	d	a	b
d	d	c	b	a

·	a	b	c	d
a	a	a	a	a
b	a	b	a	b
c	a	a	c	c
d	a	b	c	d

+	x	y	z	t
x	x	y	z	t
y	y	z	t	x
z	z	t	x	y
t	t	x	y	z

·	x	y	z	t
x	x	x	x	x
y	x	y	z	t
z	x	z	x	z
t	x	t	z	y

Solution: The rings are not isomorphic because the additive group of R is the Klein-4-group and the additive group of R' is a cyclic group of order 4.

5. For any ring R, define the opposite ring R^* as the same abelian group R with multiplication reversed, *i.e.* with multiplication defined by $a * b = b \cdot a$. Show that R^* is a ring. What is R^* when R is commutative?

Solution: The axioms for rings R^* can be verified easily. If R is commutative $R = R^*$.

6. For any abelian group G, show that the set R of all homomorphisms of G in G forms a ring for + and · defined by

$$(f + g)(a) = f(a) + g(a), \quad f, g \in R, a \in G$$

$$(f \cdot g)(a) = f(g\,(a)), \quad f, g \in R, a \in G.$$

Show that R is a ring with unit element but not commutative in general. (R is called the endomorphism ring of G).

Solution: The axioms for ring R can be verified from the axioms for the group G by evaluating on each element. R has a unit element, *viz.*, the identity map of G is G. R is not commutative.

7. Show that any ring isomorphic to a field is a field.

Solution: Let F be a field and $\theta : F \to R$ an isomorphism. The operations of + and · in R can be matched with the operations of + and · in F via the map θ, and since these operations satisfy the conditions for a field on F, the operations of $+$ and $\cdot$ in R satisfy the conditions for a field.

8. Show that any isomorphism Z onto Z is identity.

Solution: Let $\theta : Z \to Z$ be an isomorphism and let $\theta(1) = n, n \in Z$. We claim that $n = 1$. If $n \neq \pm 1$, θ cannot be surjective. Hence $n = \pm 1$. If $n = -1$, θ cannot be multiplicative, so $n = 1$ and $\theta = Id$.

9. Let R be the ring of 2×2 matrices and R^* the opposite ring of R. Define the mapping $f : R \to R^*$ by $f(A) = A'$ where if

$$A = \begin{pmatrix} a & b \\ c & d \end{pmatrix}, \text{ then } A' = \begin{pmatrix} a & c \\ b & d \end{pmatrix}.$$

Show that f is an isomorphism of R onto R^*.

Solution: Clearly $f(A + B) = f(A) + f(B)$ and $f(I) = I$.

Now $f(AB) = (AB)' = B'A' = A' * B'$ where $*$ is the multiplication in R^*. Hence f is multiplicative. Since f is also bijective, it is an isomorphism.

10. Let G and G' be isomorphic abelian groups. Show that with trivial multiplication on G and G', they are also isomorphic as rings.

Solution: Let $f : G \to G'$ be an isomorphism of abelian groups.

Then $$f(ab) = f(0) = 0 = f(a)\,f(b).$$

Hence f is an isomorphism of rings.

11. What is the endomorphism ring of the following abelian groups?

(*i*) Z (*ii*) Z_p, finite cyclic group of prime order.

Solution: (*i*) Let $f : Z \to Z$ be an endomorphism and $f(1) = n$.

Then f is completely determined by the value n. The mapping θ : End $(Z) \to Z$ given by $\theta(f) = n$ gives an isomorphism of End (Z) onto Z.

(*ii*) Let $f : Z_p \to Z_p$ be an endomorphism and $f(\bar{1}) \to \bar{i}$. Then f is completely determined by the value $\bar{i}$. The mapping θ : End $(Z_p) \to Z_p$ given by $\theta(f) = \bar{i}$ is an isomorphism of End (Z_p) onto Z_p.

2.4 HOMOMORPHISM

1. Let R be a ring with unit element e. Show that the mapping $f : Z \to R$, given by $f(n) = ne$ is a homomorphism of Z in R. When is f an embedding of Z in R?

Solution: The condition that f is a homomorphism can be verified easily. f is injective when ch. $R = 0$. In this case f is an embedding of Z in R.

2. Show that it is impossible to find a homomorphism of Z_4 onto Z_5.

Solution: Let $f : Z_4 \to Z_5$ be an "onto" homomorphism.

Then $$Z_4/\text{Ker } f \approx Z_5.$$

This is a contradiction an $Z_4/\text{Ker } f$ can have utmost 4 elements.

3. Let $f: R \to R'$ be a homomorphism of R onto R'. Show that (*i*) if R is commutative so is R', (*ii*) if R has a unit element, so has R', (*iii*) if R is an integral domain, is R' an integral domain? Why?

Solution: (*i*) Let $a', b', \in R'$. Write $a' = f(a)$ and $b' = f(b)$.

Then $$a' \cdot b' = f(a) \cdot f(b) = f(ab)$$
$$= f(ba) = f(b) \cdot f(a) = b' \cdot a'$$

Hence R' is commutative.

(*ii*) Let 1 be unit element of R and $f(1) = 1'$.

For any $a' \in R'$, write $a' = f(a)$.

Then $$1' \cdot a' = f(1) f(a) = f(1 \cdot a) = f(a) = a'$$

and $$a' \cdot 1' = f(a) \cdot f(1) = f(a \cdot 1) = f(a) = a'$$

Hence $1'$ is unit element of R'.

(*iii*) No. Take $R = Z$ and $R' = Z_n$, n not prime.

4. Let R be a ring with unit element and R' an integral domain. If $f: R \to R'$ is a non-zero homomorphism, show that $f(e)$ is the unit element of R'.

Solution: Let $f(e) = e'$. Then $e' \neq 0$, since f is a non-zero homomorphism.

Also $$e'^2 = f(e) \cdot f(e) = f(e) = e'.$$

Now for any $a' \in R'$

$$(a'e' - a')\, e' = a'\, e'^2 - a'e' = 0.$$

Since R' is a domain and $e' \neq 0$, $a'\, e' - a' = 0$.

Similarly $e'a' - a' = 0$. Hence e' is the unit element of R'.

5. Let R be a ring and X any set. Let $C(X, R)$ be the ring of all mappings of X in R. Show that for any homomorphism $f: R \to R'$ of rings the mapping $f_* : C(X, R) \to C(X, R')$ defined by $f_*(g) = f \circ g$, $g \in C(X, R)$ is a homomorphism of rings.

Solution: Clearly f_* preserves sum.

Also $$f_*(g_1 g_2) = f \circ (g_1 \cdot g_2) = (f \circ g_1) \cdot (f \circ g_2).$$

If g is the constant map mapping onto the identity of R, then $f^*(g)$ is the constant mapping onto the identity of R'. Hence f_* is a homomorphism of rings.

6. Let R be a field of ch $p > 0$. Show that the mapping $f: R \to R$ given by $f(a) = a^p$ is a ring homomorphism. Is this an isomorphism? Why?

Solution: $$f(a + b) = (a + b)^p = a^p + b^p \text{ as } p \,\Big|\, \binom{p}{i},$$

$$i \geq 1 \text{ and } \binom{p}{i} = 0.$$

Also $$f(ab) = (ab)^p = a^p b^p = f(a) \cdot f(b).$$

$f(1) = 1$. Thus f is a ring homomorphism. If $f(a) = a^p = 0$, then $a = 0$. Hence f is injective. If R is a finite field, then f is also surjective, *i.e.* f is an isomorphism. If R is not a finite field, f need not be surjective.

7. Let R_1 and R_2 be two rings, and consider the direct product $R = R_1 \times R_2$ of the abelian groups R_1 and R_2. Define a multiplication in R by

$$(a_1, a_2) \cdot (b_1, b_2) = (a_1b_1, a_2b_2),\ a_1, b_1 \in R_1 \text{ and } a_2, b_2 \in R_2.$$

Show that R is a ring. (called the direct product of R_1 and R_2).

Solution: R is clearly an abelian group as R_1 and R_2 are abelian groups. Moreover multiplication in R is associative and distributes over addition, as R_1 and R_2 are rings.

8. Let $R = R_1 \times R_2$ be direct product of rings. Show that

(*i*) if R_1 and R_2 have unit elements, so has R

(*ii*) if R_1 and R_2 are commutative, so is R

(*iii*) if R_1 and R_2 are integral domains, is R an integral domain? Why?

Solution: Let $e_1 \in R$ and $e_2 \in R_2$ be the unit elements. Then

(*i*) $(e_1, e_2) \in R$ is the unit element of R as

$$(e_1, e_2)(a, b) = (e_1a, e_2b) = (a, b)$$

and $$(a, b)(e_1, e_2) = (ae_1, be_2) = (a, b).$$

(*ii*) $$\begin{aligned}(a_1, a_2)(b_1, b_2) &= (a_1\, b_1, a_2b_2)\\ &= (b_1\, a_1, b_2a_2)\\ &= (b_1, b_2)(a_1, a_2)\end{aligned}$$

Hence R is commutative.

(*iii*) No. Let $R_1 = R_2 = Z$, $R = Z \times Z$ has zero divisors; *viz.*,

$$(2, 0)(0, 3) = (0, 0).$$

9. Let $R = R_1 \times R_2$ be direct product of rings. Show that the following mappings are all ring homomorphisms. Find the kernel in each case.

(*i*) $i_1 : R_1 \to R,\quad i_1\,(a_1) = (a_1, 0)$

(*ii*) $i_2 : R_2 \to R,\quad i_2\,(a_2) = (0, a_2)$

(*iii*) $p_1 : R \to R_1,\quad p_1\,(a_1, a_2) = a_1$

(*iv*) $p_2 : R \to R_2,\quad p_2\,(a_1, a_2) = a_2$

Solution: All the above mappings are ring homomorphisms and this can be easily verified. The kernels are (*i*) $\{0\}$, (*ii*) $\{0\}$, (*iii*) $i_2\,(R_2)$ (*iv*) $i_1\,(R_1)$.

10. Show that the ring Z_6 is isomorphic to the direct product of Z_2 and Z_3.

Solution: Follows from the next more general problem.

11. Show that if m and n are relatively prime, $Z_{mn} = Z_m \times Z_n$. Deduce that

$$U\,(Z_{mn}) = U\,(Z_m) \times U\,(Z_n).$$

where $U\,(R)$ is the group of units of R.

Solution: Since $(m, n) = 1$, there exist $\lambda, \mu \in Z$ such that $\lambda m + \mu n = 1$. Consider the map $\theta : Z_m \times Z_n \to Z_{mn}$ given by

$$\theta\,(\bar{i}, \bar{j}) = n\,\bar{i} + m\,\bar{j}\,.$$

θ is well defined and is a group homomorphism. Morover θ is "onto" because $\bar{1} = \theta(\bar{\mu}, \bar{\lambda})$. Since both groups have the same number of elements, *viz. mn*, θ is an isomorphism. In the direct product $R_1 \times R_2$, (a_1, a_2) is a unit, if and only if, both $a_1 \in R_1$ and $a_2 \in R_2$ are units.

Thus $\quad U(R_1 \times R_2) = U(R_1) \times U(R_2)$.

In particular, $U(Z_{mn}) = U(Z_m) \times U(Z_n)$ where $(m, n) = 1$.

12. Using the above exercise, show that, if $(m, n) = 1$, $\phi(mn) = \phi(m) \cdot \phi(n)$. In particular, if $n = p_1^{\alpha 1} \dots p_k^{\alpha k}$, show that

$$\phi(n) = n\left(1 - \frac{1}{p_1}\right)\left(1 - \frac{1}{p_2}\right) \dots \left(1 - \frac{1}{p_k}\right).$$

Solution: Since $U(Z_{mn}) = U(Z_m) \times U(Z_n)$ when $(m, n) = 1$ the orders on both side are the same.

Hence $\quad \phi(mn) = \phi(m)\,\phi(n)$. If $n_1 = p_1^{\alpha_1} \dots p_k^{\alpha_k}$, then

$$\phi(n) = \phi(p_1^{\alpha_1}) \dots \phi(p_k^{\alpha_k}).$$

Now
$$\phi(p_i^{\alpha_i}) = p_i^{\alpha_i} - p_i^{\alpha_i - 1} = p_i^{\alpha_i}\left(1 - \frac{1}{p_i}\right).$$

Hence
$$\varphi(n) = n\left(1 - \frac{1}{p_1}\right) \dots \left(1 - \frac{1}{p_k}\right).$$

13. Let R be a commutative ring. Show that there exists another commutative ring R' with unit element which contains R as a subring.

Solution: Let $R' = R \oplus Z$ be a direct product of abelian groups R and Z. Define a multiplication in R' by

$$(a, m) \cdot (b, n) = (ab + mb + na, mn)$$

and addition componentwise. Then R' is a ring for these operations. The mapping $\theta : R \to R'$ given by $\theta(a) = (a, 0)$ is a ring homomorphism which is injective. Hence R can be identified as a subring of R'. $(0, 1)$ is the identity element of R' as

$$(a, n)(0, 1) = (a, n) \quad \text{and} \quad (0, 1)(a, n) = (a, n).$$

14. Show that the rings Z and $Z \times Z$ are not isomorphic.

Solution: $Z \times Z$ has zero divisors, *viz.* $(n, 0)$, and $(0, m)$, $m, n \in Z$, non-zero, but Z has no zero divisors. Hence they are not isomorphic.

2.5 QUOTIENT FIELDS

1. What are quotient fields of the following integral domains?

(*i*) $R\{a + b\sqrt{-5} \mid a, b \in Z\}$.

(*ii*) $R = 2Z$

Solution: (*i*) $Q(\sqrt{-5}) = \{p + q\sqrt{-5} \mid p, q \in Q\}$

(*ii*) Q the field of rational numbers.

2. What is the quotient field of a finite integral domain?

Solution: Since a finite integral domain is a field, the quotient field coincides with itself.

3. Let F be the quotient field of R. Show that F can also be the quotient field of another integral domain R', $R \subset R'$, $R \neq R'$.

Solution: Let $R = 2Z$, $R' = Z$. Then $F = Q$ is the quotient field of both R and R'.

4. If R has a unit element 1, show that $\overline{1} = (1, 1)$ is a unit element of F, and the embedding map of R into the quotient field can be taken as $a \to \overline{(a, 1)}$.

Solution: Clearly $\overline{(a, b)}\,\overline{(1, 1)} = \overline{(a, b)} = \overline{(1, 1)}\,\overline{(a, b)}$

The embedding $f : R \to F$ is given by

$$f(a) = \overline{(ab, b)} = \overline{(a, 1)} \text{ as } (ab, b) \sim (a, 1).$$

5. If R is isomorphic to R', show that their quotient fields F and F' are isomorphic.

Solution: Let $f : R \to R'$ be an isomorphism. f defines a map $\theta : F \to F'$ by setting

$$\theta\,\overline{(a, b)} = \overline{(f(a), f(b))}$$

θ is well defined and is an isomorphism of fields.

6. Let R and R' be integral domains such that their quotient fields F and F' are isomorphic. Are R and R' isomorphic? Why?

Solution: No. Let $R = Z$ and $R' = Q$. Both have the same quotient field Q.

7. Show that an integral domain and its quotient field have the same characteristic.

Solution: Let ch $R = p \geq 0$ and F be the quotient field of R. Since R can be embedded in F, R can be considered as a subring of F. Hence, they have the same characteristic.

8. Show that a field of ch 0 contains a subfield isomorphic to Q.

Solution: Let ch $F = 0$. Consider the map $\theta : Z \to F$ defined by $\theta(1) = 1_F$ the identity element of F. θ is a ring homomorphism which is injective and can be extended to an injective homomorphism $\overline{\theta} : Q \to F$ by setting $\overline{\theta}\,(m/n) = \theta(m) \cdot \theta(n)^{-1}$, $n \neq 0$. Hence F contains $\theta(Q) \approx Q$.

9. A set M in a commutative, ring R is called a multiplicatively closed if (*i*) $0 \notin M$ and (*ii*) $ab \in M$ whenever $a \in M$ and $b \in M$. Show that if R is an integral domain, the set $M = R - \{0\}$ is multiplicatively closed set in R.

Solution: $0 \notin M$. If $a \in M$, $b \in M$ then $a \neq 0$, and $b \neq 0$, and so $ab \neq 0$. Thus $ab \in M$.

10. Let S be a multiplicatively closed set in a commutative ring with unit element and $1 \in S$. Let $F = \{(a, b) \mid a, b \in R, b \in S\}$. Define a relation $\sim$ in F by $(a, b) \sim (a', b')$ if and only if $b''(ab' - a'b) = 0$ for some $b'' \in S$. Show that $\sim$ is an equivalence relation. If R_S is the set of equivalence classes, defining + and · as in the case of quotient field construction, show that R_S is a commutative ring with unit element (R_S is called the localization of R at S).

Solution: That the relation is an equivalence relation is easy to verify. **The** operations of + and · are defined as in the quotient field construction, *viz.*

$$\overline{(a, b)} + \overline{(c, d)} = \overline{(ad + ba, bd)}$$

$$\overline{(a, b)} \cdot \overline{(c, d)} = \overline{(ac, bd)}$$

These can be verified to be well defined and R_S becomes a ring. Since R is commutative R_S is also commutative and has $\overline{(1, 1)}$ as unit element.

11. Show that the mapping $R \to R_S$ defined by $a \to \overline{(a, 1)}$ is a ring homomorphism which is injective if and only if no element of S is a zero divisor of R. **What are** the units of R_S?

Solution: Since $\overline{(a, 1)}\,\overline{(b, 1)} = \overline{(ab, 1)}$

and $\overline{(a, 1)} + \overline{(b, 1)} = \overline{(a + b, 1)}$,

the mapping is a ring homomorphism. The kernel consists of $a \in R$ with $\overline{(a, 1)} = \overline{(0, 1)}$ zero element of R_S. Thus $as = 0$ for some $s \in S$. If no element of S is a zero divisor of R, $a = 0$, *i.e.* the map is injective. Units of R_S are those $\overline{(a, b)}$ for which $a \in S$, with the inverse $\overline{(b, a)}$.

12. Obtain the quotient field of an integral domain as a special case of the ring R_S.

Solution: Let R be an integral domain and $S = R - \{0\}$. Then R_S is the quotient field of R.

2.6 IDEALS

1. Show that any non-zero ideal of Z is of the form mZ for some positive integer m.

Solution: Let I be a non-zero ideal of Z. Then I a subgroup of the abelian group Z and hence $I = mZ$, $m > 0$.

2. Show that the intersection of two ideals is an ideal.

Solution: Let I and J be ideals in R and $K = I \cap J$. Clearly K is an abelian subgroup of R. If $x \in K$, $a \in R$, $x \in I$, $x \in J$ and so $ax \in I$, $xa \in I$, $ax \in J$ $xa \in J$. Then $ax \in K$ and $xa \in K$ so that K is an ideal of R.

3. Let I and J be ideals in R. Show that $I \cup J$ is not in general an ideal in R. Show further that $I \cup J$ is an ideal if and only if either $I \subset J$ or $J \subset I$.

Solution: (*i*) Consider the example. Let $R = Z$, $I = mZ$, $J = nZ$, $m \neq n$. m, n not multiples of each other.

Then $I \cup J$ is not an ideal as $m - n \notin I \cup J$.

(*ii*) If either $I \subset J$ or $J \subset I$, then $I \cup J = I$ or $I \cup J = J$ and hence it is an ideal. Conversely let $I \cup J$ be an ideal. If $I \not\subset J$, then there exists $a \in I$, $a \notin J$. Also if $J \not\subset I$, there exists $b \in J$, $b \notin I$. Then $a + b \notin I$ as $a \in I$, $b \notin I$. Also $a + b \notin J$ as $b \in J$, $a \notin J$. Thus $a + b \notin I \cup J$, a contradiction.

4. Let F be a field. Show that any homomorphism of F in F is either a trivial map or an isomorphism of F in F.

Solution: Let $\theta : F \to F$ be a homomorphism and $K = \text{Ker } \theta$. Since K is an ideal of F, $K = 0$ or $K = F$. If $K = 0$, θ is an isomorphism of F in F. If $K = F$, θ is trivial.

5. Let I be an ideal in a commutative ring. Show that $R : I = \{a \in R \mid aR \subset I\}$ is an ideal in R (called the quotient ideal of I in R).

Solution: The conditions for the ideal can be verified easily as I is an ideal.

6. Let R be a commutative ring. Show that $N = \{a \mid a \in R,\ a^m = 0 \text{ for some } m \geq 1\}$ is an ideal in R. (called the nil radical of R).

Solution: Let $a \in N$ and $b \in N$ so that $a^m = 0$, $b^n = 0$. Then

$$(a + b)^{m+n} = \sum_{i=0}^{m+n} \binom{m+n}{i} a^i\, b^{m+n-i}$$

The expression on the right is zero as either $i \geq m$ or $m + n - i \geq n$. Thus $a + b \in N$. Also for any $x \in R$

$$(xa)^m = x^m\, a^m = 0 \text{ and } (ax)^m = a^m\, x^m = 0.$$

Thus N is an ideal in R.

7. Let R be a commutative ring and I an ideal in R. Show that

$$\sqrt{I} = \{a \in R \mid a^n \in I \text{ for some } n \geq 1\}$$

is an ideal in R (called the radical of I). What is $\sqrt{I}$ when $I = \{0\}$?

Solution: If $a \in \sqrt{I}$ and $b \in \sqrt{I}$, then $a^m \in I$, $b^n \in I$ for some $m, n \geq 1$.

$$(a + b)^{m+n} = \sum_{i=0}^{m+n} \binom{m+n}{i} a^i\, b^{m+n-i}$$

The expression on the right belongs to I as either $i \geq m$ or $m + n - i \geq n$. The $a + b \in \sqrt{I}$. Also xa, $ax \in \sqrt{I}$, for $x \in R$. If $I = 0$, $\sqrt{I}$ is the nil radical of R.

8. Let R be a ring and $a \in R$. Show that $I = \{x \mid x \in R,\ xa = 0\}$ is a left ideal of R (called left annihilator of a) and $J = \{x \mid x \in R,\ ax = 0\}$ is a right ideal of R (called right annihilator of a).

Solution: $xa = 0$ and $ya = 0$ implies $(x - y)\, a = 0$ and $(bx)\, a = b\, (xa) = 0\ b \in R$. Thus I is a left ideal. Similarly J is a right ideal.

9. Let I be an ideal in a ring R and J an ideal in the ring I. Show by an example that J need not be an ideal in R.

Solution: Let $R = M_n(Z)$ and $I = \left\{\begin{pmatrix} a & 0 \\ b & 0 \end{pmatrix} \middle| \, a, b \in Z\right\}$.

Then I is a left ideal of R.

$$\text{Let } J = \left\{\begin{pmatrix} 0 & 0 \\ d & 0 \end{pmatrix} \middle| \, d \in Z\right\}.$$

Then J is a left ideal in I. But J is not a left ideal in R because

$$\begin{pmatrix} 0 & 1 \\ 0 & 1 \end{pmatrix}\begin{pmatrix} 0 & 0 \\ 1 & 0 \end{pmatrix} = \begin{pmatrix} 1 & 0 \\ 1 & 0 \end{pmatrix} \notin J.$$

10. Find all ideals of the direct product ring $Z \times Z$.

Solution: Let I be an ideal of $Z \times Z$. If $(a, b) \in I$, then $(1, 0)(a, b) = (a, 0) \in I$ and $(a, b)(0, 1) = (0, b) \in I$. Thus I is generated by elements of the type $(a, 0)$, $(0, b)$. Consider

$$J = \{a \mid (a, 0) \in I\} \text{ and } K = \{b \mid (0, b) \in I\}.$$

Then J and K are ideal of Z and $I = J \times K$ as $(a, b) \in I$ can be written as $(a, b) = (a, 0) + (0, b)$.

Then any ideal of $Z \times Z$ is of the form $mZ \times nZ$, $m, n \in Z$.

11. Let R be a commutative ring, $a_1, ..., a_n \in R$. Show that $I = Ra_1 + ... + Ra_n$ is an ideal in R.

Solution: If $x, y \in I$, then $x = \lambda_1 a_1 + ... + \lambda_n a_n$

and $\quad y = \mu_1 a_1 + ... + \mu_n a_n.$

Hence $\quad x - y = (\lambda_1 - \mu_1) a_1 + ... + (\lambda_n - \mu_n) a_n \in I$

and $\quad ax = a\lambda_1 a_1 + ... + a\lambda_n a_n \in I \; a \in R.$

Hence I is an ideal.

12. Given any ideal I in a commutative ring R, can you always find $a_1, ..., a_n \in I$ such that $I = Ra_1 + ... + Ra_n$? Why?

Solution: No. Only ideals which are finitely generated have this property. There are rings which contain non-finitely generated ideals. (Exercise 12, 2.12)

13. Prove or disprove : For ideals I, J, K, $I(J + K) = IJ + IK$.

Solution: True. Any element of $I(J + K)$ is of the form $\sum_i a_i (b_i + c_i)$, $a_i \in I$, $b_i \in J$, $c_i \in K$. This is the same as

$$\sum_i a_i b_i + \sum_i a_i c_i \in IJ + IK.$$

Conversely $\quad IJ \subset I(J + K)$ as $J \subset J + K$

and $\quad IK \subset I(J + K)$ as $K \subset J + K$.

Hence $\quad IJ + IK \subset I(J + K)$.

Thus $\quad IJ + IK = I(J + K)$.

14. Show that the nil radical of Z_n is non-zero, if and only if, n is divisible by the square of a prime.

Solution: Let n be divisible by the square of a prime p.

Then $$n = p^{\lambda} m, \lambda \geq 2, (m, p) = 1.$$

Then $$\bar{a} = \overline{pm} \in Z_n \; \bar{a} \neq \bar{0} \text{ and } \bar{a}^{\lambda} = 0.$$

Thus Z_n has non-zero element $\bar{a}$ in the nil radical. Conversely assume that there exist $\bar{a} \neq 0$ in the nil radical. Then $\bar{a}^{\lambda} = \bar{0}$ $(\lambda \geq 2)$. If p is a prime factor of a, p^{λ} divides n, $\lambda \geq 2$.

2.7 QUOTIENT RINGS

1. Find all the quotient rings of Z.

Solution: Any quotient ring of Z is of the type Z/nZ, $n \geq 0$.

2. Let $f : Z \to F$ be a homomorphism of Z onto F where F is a field. Show that F must be a field with prime number of elements.

Solution: Let $K = \text{Ker} f$. Then $K = nZ$, $n \geq 1$. Now $F \approx Z/nZ$ and since F is a field, n is a prime, *i.e.* $F \approx Z_p$.

3. If R is a commutative ring with unit element, show that the same is true of R/I. Is the converse true? Why?

Solution: R/I is commutative because R is commutative. If $1 \in R$ is the unit element, $\bar{1} \in R/I$ is the unit element of R/I. Converse is not true as the following example shows.

Example: Let R be a commutative ring and R' a non-commutative ring. The direct product ring $R \times R'$ is not commutative. It has an ideal $\{0\} \times R'$ whose quotient is isomorphic to R, and hence commutative.

4. Write down all homomorphic images of Z_{15}.

Solution: The ideals of Z_{15} and $\{\bar{0}\}$, $\{\bar{3}\}$, $\{\bar{5}\}$ and Z_{15}. The corresponding quotient rings are isomorphic to Z_{15}, Z_3, Z_5 and $\{0\}$.

5. Let N be the nil radical of R. Show that R/N has non-zero nilpotent elements.

Solution: Let $\bar{a} \in R/N$ be nilpotent so that $\bar{a}^n = 0$ for some $n \geq 1$. Then $a^n \in N$ and since N is nilpotent, $(a^n)^m = 0$, $a^{mn} = 0$. Hence $a \in N$ *i.e.* $\bar{a} = 0$.

6. Let $n = mk$ and let I denote the set of all multiples of $\bar{m}$ in Z_n. Show that $Z_n/I \approx Z_m$.

Solution: Consider the map $Z_n \to Z_m$ given by $i(\text{mod } n) \to i(\text{mod } m)$. The mapping is well defined as $m \mid n$ and is a homomorphism of Z_n onto Z_m. If has kernel = $(\bar{m})$.

7. Let $M_2(R)$ be the ring of all 2×2 matrices with coefficients in a ring R. Show that if $f : R \to R'$ is a ring homomorphism, the mapping $\bar{f} : M_2(R) \to M_2(R')$

given by $\overline{f}(A) = A'$, where

$$A' = \begin{pmatrix} f(a) & f(b) \\ f(c) & f(d) \end{pmatrix}, \text{ if } A = \begin{pmatrix} a & b \\ c & d \end{pmatrix}, \text{ is also a ring homomorphism.}$$

Solution: Let $B = \begin{pmatrix} a' & b' \\ c' & d' \end{pmatrix}$. Then

$$f(A+B) = \begin{pmatrix} f(a+a') & f(b+b') \\ f(c+c') & f(d+d') \end{pmatrix}$$

$$= \begin{pmatrix} f(a)+f(a') & f(b)+f(b') \\ f(c)+f(c') & f(d)+f(d') \end{pmatrix} = f(A)+f(B)$$

Also $\quad f(AB) = f\begin{pmatrix} aa'+bc' & ab'+bd' \\ ca'+dc' & cb'+dd' \end{pmatrix}$

$$= \begin{pmatrix} f(a)f(a')+f(b)f(c') & f(a)f(b')+f(b)f(d') \\ f(c)f(a')+f(d)f(c') & f(c)f(b')+f(d)f(d') \end{pmatrix}$$

$$= f(A)f(B)$$

$$f(Id) = Id$$

Hence f is a ring homomorphism.

8. Let $f : R \to R'$ be a ring homomorphism of R in R' and let J and J' be ideals in R and R', respectively, such that $f(J) \subset f(J')$. Show that f defines a ring homomorphism $\overline{f} : R/J \to R'/J'$. What is the kernel of $\overline{f}$?

Solution: The mapping $\overline{f}(a + J) = f(a) + J'$ is well defined because $f(J) \subset J'$. Moreover it preserves addition and multiplication. Thus $\overline{f}$ is a ring homomorphism

$$\text{Ker } \overline{f} = \frac{\text{Ker}(f) + J}{J}.$$

9. Let $f : R \to R'$ be a ring homomorphism of R onto R' and J' an ideal in R', $f^{-1}(J') = J$. Show that J is an ideal in R and $R/J \simeq R'/J'$. Deduce the following isomorphism theorem. If I and J are ideals in R such that $J \subset I$, then the quotient ring $\dfrac{(R/J)}{(I/J)}$ is isomorphic to R/I.

Solution: That J is an ideal R can be easily verified using the fact that J' is an ideal in R'. The mapping $R \xrightarrow{f} R' \xrightarrow{p} R'/J'$ given by pof is a homomorphism "onto" with kernel equal to J. Thus $R/J \approx R'/J'$. In particular let $R' = R/J$ and $f : R \to R'$ the natural projection. Let $J' = I/J$. Then $R'/J' = \dfrac{R/J}{I/J} \approx R/I$.

10. Prove the following isomorphism theorem for rings : If I and J are ideals in R, then the quotient ring $\frac{(I+J)}{J}$ is isomorphic to $I/I \cap J$.

Solution: Consider the projection map $p : I \to \frac{I+J}{J}$ given by $p(a) = a + J$. p is a ring homomorphism, "onto" with $\text{Ker } p = I \cap J$. Hence $I/I \cap J \approx \frac{I+J}{J}$.

11. Let R be a non-trivial ring with 1 in which every subring is an ideal. Show that $R \approx Z$ or Z_n for some n.

Solution: Consider the map $f : Z \to R$ given by $f(1) = e \in R$ the identity element. f is a ring homomorphism. If $\text{Ker} f = 0$, *i.e.* ch $R = 0$, then $Z \approx f(Z) \subset R$ which is an ideal by assumption. Since $f(Z)$ contains unit element of R, $f(Z) = R$. Hence $Z \approx R$. If $\text{Ker} f \neq 0$, $\text{Ker} f = nZ$. Then $Z_n \approx f(Z_n) \subset R$ and since $f(Z_n)$ contains the unit element of R and is an ideal, $f(Z_n) = R$. Hence $Z_n \approx R$.

12. Let I be an ideal in R and S a subring such that $I \cap S = \{0\}$. Show that R/I contains a subring isomorphic to S.

Solution: Consider the projection map $p : R \to R/I$ and its restriction to S. It has kernel $S \cap I = \{0\}$. Hence it is an isomorphism of S onto its image which is a subring of R/I.

13. Show that any field F of ch $p > 0$ contains a subfield isomorphic to Z_p.

Solution: Consider the map $f : Z \to F$ given by $f(1) = 1_F$. f is a ring homomorphism with $\text{Ker} f = pZ$, p prime. Hence Z_p is isomorphic to its image, a subfield of F.

14. A ring R is called simple if its only ideals (two-sided) are 0 and R. Show that the characteristic of a simple ring is either 0 or a prime.

Solution: Let $n = \text{ch } R$, and $n \neq 0$. Suppose n is composite. Then n has a proper divisor d such that $n = dd'$, where d' is also a proper divisor of n. Now $d'y$ cannot be zero for all y, since $d' < n$ and $n = \text{ch } R$. So there exists some y with $z = d'y \neq 0$.

Now $\qquad dz = dd'y = ny = 0.$

Consider $\qquad I = \{x \mid x \in R, dx = 0\}.$

Then $\qquad I \neq 0$ as $z \in I$, $z \neq 0$.

Also $I \neq R$ since $d < n = \text{ch } R$. Also I is a two sided ideal of R, as $d(ax) = a(dx) = 0$, $a \in R$ contradicting the fact that R is a simple ring.

2.8 PRIME AND MAXIMAL IDEALS

1. Investigate whether the following ideals are (*i*) prime ideals, (*ii*) maximal ideals in R.

(*i*) $R = Z_6$ and $I = \{\overline{0}, \overline{2}, \overline{4}\}$.

(*ii*) $R = 2Z$ and $I = 4Z$.

(*iii*) $R = C[0, 1]$ the ring of continuous functions on $[0, 1]$ and $I = \{x \mid x(1/2) = 0\}$.

Solution: (*i*) I is a maximal ideal of R as $R/I \approx Z_2$ is a field. Hence, it is a also a prime ideal.

(*ii*) I is a maximal ideal of R. If $J = nZ \supset I$, 2 divides n which divides 4. Hence $n = 2$ or $n = 4$ *i.e.* $J = R$ or $J = I$. It is also a prime ideal.

(*iii*) I is a prime ideal of R. Consider the map $\theta : C[0, 1] \to F$. F reals defined by $\theta(x) = x(1/2)$. It is a ring homomorphism with Ker $\theta = I$. Hence $\frac{C[0,1]}{I} \approx \text{Im}(\theta)$. Also θ is "onto" because given any value λ of F, the exists a continues map x on $[0, 1]$ with $x(1/2) = \lambda$. Since F is a field, I is a maximal ideal, and also a prime ideal.

2. Find all the prime ideals and maximal ideals of Z_{16}.

Solution: Any ideal $I \supset 16Z$ is of the form nZ, where $n \mid 16$. Hence $n = 1, 2, 4, 8, 16$. Hence, the ideal of Z_{16} are Z_{16}, $(\bar{2})$, $(\bar{4})$, $(\bar{8})$ and $\{\bar{0}\}$. $(\bar{2})$ is a maximal ideal as $Z_{16}/(\bar{2}) \approx Z_2$ a field. $(\bar{4})$ is not prime because $(\bar{4})^2 = 0$, $\bar{4} \neq 0$. $I = (\bar{8})$ is also not prime as $\bar{2} \notin I$, $\bar{4} \notin I$ but $\bar{2} \cdot \bar{4} \in I$. Since Z_{16} is not a domain $\{\bar{0}\}$ is also not prime.

3. Given an example of a prime ideal in $Z \times Z$ which is not a maximal ideal.

$P = Z \times \{0\}$ is an ideal in $R = Z \times Z$ and $R/P \approx \{0\} \times Z \approx Z$ is an integral domain. Hence P is a prime ideal which is not maximal.

4. Let $f : R \to R'$ be a ring homomorphism of R in R'. Let I be a prime (maximal) ideal in R. Is $f(I) = I'$ is a prime (maximal) ideal in R'? If so why?

Solution:

(*i*) No. The projection map $p : Z \to Z_n$ (n not prime) maps the zero ideal onto zero ideal which is not prime.

(*ii*) No. The map $p : Z \to Z_4$ maps the maximal ideal $3Z$ onto the whole of Z_4 which is not a maximal ideal of Z_4 being improper ideal.

5. Let $f : R \to R'$ be a ring homomorphism of R in R'. Let I' be a prime (maximal) ideal of R' and let $I = f^{-1}(I')$. Can you say that I is a prime (maximal) ideal of R? If so why?

Solution: (*i*) Yes for prime ideals. If $ab \in I = f^{-1}(I')$ then $f(ab) \in I'$, *i.e.* $f(a) f(b) \in I'$. Hence $f(a) \in I'$ or $f(b) \in I'$, *i.e.* $a \in I$ or $b \in I$.

(*ii*) No for maximal ideals. Consider the inclusion map $i : Z \to Q$. $\{0\}$ is a maximal ideal of Q but $i^{-1}(\{0\}) = \{0\}$ is not a maximal ideal of Z.

6. Show that P is a prime ideal of R if and only if $R - P$ is a multiplicatively closed set.

Solution: Let $S = R - P$. If $a, b \in S$, then $a \notin P$ and $b \notin P$ and so $ab \notin P$, *i.e.* $ab \in S$. Converse is similar.

7. Let R be a commutative ring with unit element in which the set of non-regular elements forms an ideal M. Show that M is a maximal ideal of R and that M contains any other proper ideal of R. (Such a ring R is called a local ring).

Solution:

(*i*) Let I be an ideal properly containing M. Then I contains a regular element and hence $I = R$. Thus M is a maximal ideal.

(*ii*) Let I be a proper ideal of R. Then every element of I must be non-regular. Hence $I \subset M$.

8. If P is a prime ideal of R and $S = R - P$, show that R_S is a local ring.

Solution: $\overline{(a, b)} \in R_S$ is regular if and only if $a \in S = R - P$. Thus the regular elements consists of all $\overline{(a, b)} \in R_S$ with $a \in P$. These form an ideal in R_S as P is an ideal in R.

9. An ideal I in a commutative ring is called primary if, whenever $ab \in I$, either $a \in I$ or $b^n \in I$ for some $n \geq 1$. Show that

(*i*) Every prime ideal is primary.

(*ii*) A primary ideal need not be a prime ideal.

(*iii*) If I is a primary ideal, its radical $\sqrt{I}$ is a prime.

Solution:

(*i*) If P is a prime ideal and $ab \in P$, then either $a \in P$ or $b \in P$. Thus I is primary.

(*ii*) Let $I = P^2$, where P is a prime ideal with $P^2 \neq P$. Then I is a non-prime primary ideal for if $ab \in I = P^2 \subset P$, either $a \in P$ or $b \in P$. So $a^2 \in P^2 = I$ or $b^2 \in P^2 = I$.

(*iii*) Let I be a primary ideal and $\sqrt{I} = \{a \mid a^n \in I \text{ for some } n \geq 1\}$. Now $\sqrt{I}$ is a prime ideal for let $ab \in \sqrt{I}$. Then $(ab)^n = a^n b^n \in I$. Then either $a^n \in I$ or $(b^n)^m \in I$ *i.e.* $a \in \sqrt{I}$ or $b \in \sqrt{I}$.

10. Let M_1 and M_2 be distinct maximal ideals of R. Show that $R = M_1 + M_2$. Define a suitable ring homomorphism of R onto $R/M_1 \times R/M_2$ and deduce that $R/M_1 \cap M_2 \approx R/M_1 \times R/M_2$.

Solution: $M_1 + M_2 \supset M_1$ properly as M_1 and M_2 are distinct. Hence $M_1 + M_2 = R$. Define a mapping $f: R \to R/M_1 \times R/M_2$ by setting $f(a) = (a + M_1, a + M_2)$. Then f is a ring homomorphism. f is also "onto" for let $(a + M_1, b + M_2) \in R/M_1 \times R/M_2$. Write $1 = m_1 + m_2$, $m_1 \in M_1$ and $M_2 \in M_2$. Then $am_2 + bm_1$ maps onto this element as

$$am_2 + bm_1 + M_1 = a(1 - m_1) + bm_1 + M_1 = a + M_1$$

and $$am_2 + bm_1 + M_2 = am_2 + b(1 - m_2) + M_2 = b + M_2$$

The map has kernel $M_1 \cap M_2$ so that

$$\frac{R}{M_1 \cap M_2} \approx \frac{R}{M_1} \times \frac{R}{M_2}$$

11. Extend the above isomorphism to a fintie number of maximal ideals $M_1, \ldots, M_k$, with $M_i + M_j = R$, $i \neq j$ and deduce Chinese remainder theorem: Given prime $p_1, \ldots, p_k$ and integers $a_1, \ldots, a_k$, there exists an integer a such that $a \equiv a_i \pmod{p_i}$, $i \leq i \leq k$.

Solution:

(*i*) Consider the map

$$f: R \to R/M_1 \times R/M_2 \times \dots \times R/M_k$$

given by $$f(a) = (a + M_1, \dots, a + M_k).$$

Then f is a ring homomorphism. To show that f is "onto" it is enough to show that $(\overline{1}, \overline{0}, \dots, \overline{0}), (\overline{0}, \overline{1}, \dots, \overline{0}), \dots (\overline{0}, \overline{0}, \dots, \overline{1})$ all belong to the image. Consider $(\overline{0}, \overline{0}, \dots, \overline{1}, \dots \overline{0})$ with 1 in the ith place. Since $M_i + M_j = R, i \neq j$, write $1 = m_i + m_j$, $m_i \in M_i, m_j \in M_j, j \neq i$. Then $1 - m_i$ gets mapped onto $(\overline{0}, \overline{0}, \dots, \overline{1}, \dots \overline{0})$, because

$$1 - m_i = m_j \in M_j, \quad j \neq i$$

and $$1 - m_i + M_i = 1 + M_i = \overline{1}.$$

(*ii*) Consider the special case when $R = Z$ and $M_i = p_iZ$. Since p_i's are distinct primes. $M_i + M_j = 1, i \neq j$. By the above theorem $(\overline{a_1}, \overline{a_2}, \dots, \overline{a_k})$ has a premage $a \in Z$, *i.e.* $a \equiv a_i \pmod{p_i}$, $1 \leq i \leq k$.

(*iii*) The kernel of the map in (*i*) is $\bigcap_{i=1}^{k} M_i$.

Hence $$R/\bigcap_{i=1}^{k} M_i \approx \prod_{i=1}^{k} \frac{R}{M_i}.$$

12. If I is a primary ideal in a commutative ring, show that every zero divisor is nilpotent in R/I.

Solution: Let $\overline{a} \in R/I$ be a zero divisor, $\overline{a} \neq \overline{0}$. Then there exits $\overline{b} \in R/I$, $\overline{b} \neq 0$ such that $\overline{a} \cdot \overline{b} = \overline{0}$, *i.e.* $ab \in I$. Hence either $b \in I$ or $a^n \in I$. Since $\overline{b} \neq \overline{0}$, we have, $a^n \in I$, $\overline{a}^n = 0$. Hence $\overline{a}$ is nilpotent.

13. Show that for the ring R_S, $S = R - P$, P being a prime ideal in R, there is a (1, 1) correspondence between prime ideals of R_S and the prime ideals of R not intersecting S.

Solution: Consider the map $f: R \to R_S$ defined by $f(a) = \overline{(a, 1)}$. Now $\overline{(a, 1)} \in R_S$ is a unit, if and only if, $a \in S = R - P$, *i.e.* $a \notin P$. If P' is a prime ideal not intersecting S, $f(P')$ will contain only non-units and will be a proper ideal and also a prime ideal. Conversely if Q is a prime ideal of R_S and $f^{-1}(Q) = P'$, P' is a prime ideal of R not intersecting S. The bijection is given by the correspondence $P' \to P'_S = f(P')$.

14. Let P_1 and P_2 be prime ideals in a commutative ring R. Show that $P_1 \cap P_2$ is prime implies $P_1 \subset P_2$ or $P_2 \subset P_1$.

Solution: If $P_1 \subset P_2$ the result is proved otherwise we will show that $P_2 \subset P_1$. Choose $a \in P_1$, $a \notin P_2$. Let $b \in P_2$. Then $ab \in P_1 \cap P_2$ and since $P_1 \cap P_2$ is prime $a \in P_1 \cap P_2$ or $b \in P_1 \cap P_2$. Since $a \notin P_2$ we have $b \in P_1 \cap P_2 \subset P_1$. Hence, $P_2 \subset P_1$. This proves the result.

2.9 FACTORISATION

1. In the ring $R = \{a + b\sqrt{-5} \mid a, b \in Z\}$, investigate whether the following elements are irreducible.

(*i*) $3 + \sqrt{-5}$, (*ii*) $4 + 3\sqrt{-5}$.

Solution:

(*i*) Let $3 + \sqrt{-5} = \alpha\beta$

$$N(3 + \sqrt{-5}) = 3^2 + 5 \cdot 1^2 = 14 = N(\alpha)\, N(\beta).$$

Thus $N(\alpha) = 1, 2, 7$ or 14. $N(\alpha) = 2$ is impossible.

$N(\alpha) = 7$ is also impossible. Hence $N(\alpha) = 1$ or 14. Thus α is either a unit or an associate. Hence $3 + \sqrt{-5}$ is irreducible.

(*ii*) $N(4 + 3\sqrt{-5}) = 4^2 + 5 \cdot 3^2 = 16 + 45 = 61.$

If α is a divisor of $4 + 3\sqrt{-5}$ then $N(\alpha) = 1, 61$, *i.e.* α is unit or an associate.

Hence $4 + 3\sqrt{-5}$ is irreducible.

2. Show that if, $a, b \in R$, then (*i*) a divides b if, and only if, $Ra \supset Rb$, (*ii*) a is an associate of b if, and only if, $Ra = Rb$.

Solution: $Ra \supset Rb$ if, and only if, $b \in Ra$, *i.e.* $b = \lambda a$, $\lambda \in R$, *i.e.* a divides b. $Ra = Rb$ if, and only if, $Ra \supset Rb$ and $Rb \supset Ra$, *i.e.* a and b divide each other. Hence they are associates.

3. Show that if $a \in R$ is such that Ra is a maximal ideal, then a is an irreducible element.

Solution: Let b be a divisor of a so that $Rb \supset Ra$. Since Ra is maximal $Rb = R$ or $Rb = Ra$ *i.e.* b is a unit or an associate of a. Hence a is irreducible.

4. Show that any associate of an irreducible element is irreducible.

Solution: Let a be irreducible and b an associate of a. Then $a = ub$, u unit. If b has a proper divisor, it is also a proper divisor of a, a contradiction.

5. Show that in the ring $R = \{a + b\sqrt{-5} \mid a, b \in Z\}$, the elements $\alpha = 3$, and $\beta = 1 + 2\sqrt{-5}$ are relatively prime, but $\alpha\gamma$ and $\beta\gamma$ have no g.c.d. in R, where $\gamma = 7(1 + 2\sqrt{-5})$.

Solution: (*i*) $N(\alpha) = 9$ and $N(\beta) = 21$. If δ is a divisor of α and β, $N(\delta)$ divides 9 and 21. Hence $N(\delta) = 1$ or $N(\delta) = 3$. But $N(\delta) = 3$ is impossible, so that $N(\delta) = 1$, *i.e.* δ is a unit. Hence α and β are relatively prime.

(*ii*) Suppose $\alpha\gamma$ and $\beta\gamma$ has a g.c.d. δ. Since γ dives $\alpha\gamma$ and $\beta\gamma$, $\gamma \mid \delta$. Then the g.c.d of α and β is δ/γ and so $\delta/\gamma = 1$, *i.e.* $\gamma = \delta$.

Since $\alpha\gamma = 21. (1 + 2\sqrt{-5}) = (1 - 2\sqrt{-5})(1 + 2\sqrt{-5})^2$

β^2 is a common divisor of $\alpha\gamma$ and $\beta\gamma$ and so $\beta^2 \mid \delta$ and $\delta = \gamma$. This is a contradiction as $N(\beta)^2$ does not divide $N(\gamma)$.

Define the 1.c.m. of two elements $a \in R$, $b \in R$ as an element $\in R$ which satifies

(*i*) $a \mid l$ and $b \mid l$,

(*ii*) If $a \mid l'$ and $b \mid l'$, then $l \mid l'$.

6. Show that if 1.c.m. exists it is unique up to units.

Solution: Let l_1 and l_2 be two 1.c.m's of a and b. Then conditions (*i*) for l_1 and (*ii*) with l_2 implies $l_2 \mid l_2$. Similarly $l_2 \mid l_1$. Hence l_1 and l_2 differ by a unit.

7. Formulate a definition of g.c.d. and 1.c.m. for a finite number of elements.

Solution: Let $a_1, a_2, ..., a_n \in R$. An element $d \in R$ is called g.c.d. if

(*i*) $d \mid a_1, d \mid a_2, ... d \mid a_n$

(*ii*) If $d' \mid a_1, ..., d' \mid a_n$, then $d' \mid d$.

An element $l \in R$ is called an 1.c.m. if

(*i*) $a_1 \mid l, a_2 \mid l, ... a_n \mid l$

(*ii*) If $a_1 \mid l', a_2 \mid l', ... a_n \mid l'$, then $l \mid l'$.

8. Show that if $p \in R$ is a prime element such that $p \mid a_1 ... a_n$, the $p \mid a_i$ for some i.

Solution: By definition, p is prime if $p \mid a_1 a_2$ implies $p \mid a_1$ or $p \mid a_2$. The result now follows by induction on n.

9. Show that in the ring Z, $a \in Z$, $a \neq 0$, the following are equivalent: (*i*) a is prime, (*ii*) a is irreducible, (*iii*) Za is a prime ideal, (*iv*) Za is a maximal ideal.

Solution: Clearly (*i*) and (*iii*) are equivalent.

Also (*i*) $\Rightarrow$ (*ii*). Also by Problem 3, (*iv*) $\Rightarrow$ (*ii*). Also in Z, n is irreducible if, and only if, its divisors are $\pm n$ and ± 1, *i.e.* n is prime, *i.e.* (*ii*) $\Rightarrow$ (*i*). Also (*ii*) $\Rightarrow$ (*iv*) in Z. Thus all the four are equivalent.

10. Show that if $a, b \in R$, such that $ka + lb = 1$, for some $k, l \in R$, then a and b are relatively prime.

Solution: Let d be a common divisor of a and b so that d divides $ka + lb = 1$. Hence d is a unit, *i.e.* a and b are relatively prime.

2.10 EUCLIDEAN DOMAINS

1. Show that the units in the ring $Z + iZ$ of Gaussian integers are ± 1, and $\pm i$.

Solution: Let $a + ib = \alpha$ be a unit. Then $N(\alpha) = a^2 + b^2$ must divide $N(1) = 1$. So that $a^2 + b^2 = 1$. Hence $a = \pm 1$, $b = 0$ or $a = 0$, $b = \pm 1$. Hence the units are ± 1 or $\pm i$.

2. Find the g.c.d. of $2 + 3i$ and $4 + 7i$ in $Z + iZ$.

Solution:

$$4 + 7i = 2(2 + 3i) + i$$

$$2 + 3i = i(3 - 2i)$$

Hence, the g.c.d. is i.

3. Show that any non-zero ideal in $Z + iZ$ contains a positive integer.

Solution: Let I be an ideal containing a non-zero element $a + ib$. The $(a - ib)$ $(a + ib) = a^2 + b^2 \in I$ which is a positive integer.

4. Show that the ring of even integers is not a Euclidean domain.

Solution: It does not have a unit element.

5. Show that every field R is a Euclidean domain.

Solution: Define $d(a) = 1$, for each $a \neq 0$. Conditions (*i*) and (*ii*) of Euclidean domain are satisfied. As for (*iii*), for any $b \neq 0$, $a = (ab^{-1})\, b + r$, with $r = 0$. So R is a Euclidean domain.

6. Show that if a is an irreducible element in a Euclidean domain, Ra is a maximal ideal.

Solution: Let $I \supset Ra$ be any ideal. Then $I = Rb$ for some $b \in R$. Since $Rb \supset Ra$, b divides a. Hence either b is a unit or an associate of a *i.e.* $Rb = R$ or $Rb = I$. Thus I is maximal.

7. Show that, $1 + i$ is irreducible in $R = Z + iZ$. What is the field $R/R\,(1 + i)$?

Solution: Let α be a divisor of $1 + i$ so that $N(\alpha)$ divides $N(1 + i) = 2$. Hence $N(\alpha) = 1$ or 2, *i.e.* $a^2 + b^2 = 1$ or 2. Then $a = \pm 1$, $b = 0$, $a = 0$, $b = \pm 1$ or $a = \pm 1$, $b = \pm 1$. Thus, the only divisors of $1 + i$ are units and associates. Then $1 + i$ is irreducible and $R(1 + i)$ is a maximal ideal. Hence $R/R\,(1 + i)$ is a field. This field is isomorphic to $Q + iQ$.

8. Show that $R = \{a + b\sqrt{-5} \mid a, b \in Z\}$ is not a Euclidean domain.

Solution: There exists pairs of elements in this ring which not have a g.c.d. Hence, it cannot be Euclidean domain. (Problem 5, Section 2.9)

Let R be a Euclidean domain and let.

$$I = \{a \mid d(a) > d(1)\} \cup \{0\}.$$

9. Is I an ideal in R? Why?

Solution: No. Let $R = Z$. Then I consists of all non-units together with 0, *i.e.* $I = \{n \mid n \in Z, n \neq \pm 1\}$. Thus I is not an ideal in Z.

10. Show that any two elements a, b of a Euclidean domain have an 1.c.m. $[a, b]$. If (a, b) denotes the g.c.d. of a and b, show that $ab = (a, b)\,[a, b]$.

Solution: Let $I = Ra$ and $J = Rb$. Then $I \cap J$ is an ideal in R and hence $I \cap J = Rl$ for some l. Clearly $a \mid l$ and $b \mid l$ and if $a \mid l'$, $b \mid l'$ then $l \mid l'$. Thus l is the 1.c.m. of a and b. A Euclidean domain is a unique factorisation domain and we will show later that in a *UFD*, $(a, b)\,[a, b] = a \cdot b$.

11. Let (R, d) be a Euclidean domain. Define another function d' in R by $d'(a) = d(a) + m$, where m is a fixed positive integer with $d(1) + m > 0$. Show that (R, d') is a Euclidean domain.

Solution:

(*i*) $d'(a) = d(a) + m = d(1a) + m \geq d(1) + m \geq 0.$

(*ii*) $d'(ab) = d(ab) + m \geq d(a) + m = d'(a).$

For any $a, b \in R, b \neq 0, a = bq + r$

with either $r = 0$ or $d(r) < d(b)$

So $r = 0$ or $d'(r) = d(r) + m < d(b) + m = d'(b)$

Hence (R, d') is a Euclidean domain.

12. If a/b and $d(a) = d(b)$, then show that a and b are associates.

Solution: If a is a proper divisor of b, then $d(a) < d(b)$. Since $d(a) = d(b)$, a must be an associate of b.

13. Let $R = \{a/b \mid b = p^s, s, a \in Z\}$, p fixed prime. Show that R is a subring of Q. For any $q \in R, q \neq 0$, write $q = ap^{-k}$, $(p, a) = 1$, define $d(q) = |a|$. Show that (R, d) is a Euclidean domain.

Solution: Clearly elements of R are closed for addition and multiplication of Q.

Thus R is a subring of Q. Now d satisfies the axioms for Euclidean domain. Let $q = a/p^k$, $(p, a) = 1$, $k \in Z$.

(i) $d(q) = |a| \geq 0$.

(ii) If $q' = a'/p^{k'}$, then $qq' = aa'/p^{k+k'}$, $(p, aa') = 1$ and so $d(qq') = |aa'| = |a| \cdot |a'| = d(q)\, d(q') \geq d(q)$.

(iii) Let $q, q' \in R$ with $q' \neq 0$.

Write $q = a/p^k$ and $q' = b/p^l$, $(a, p) = 1$, $(b, p) = 1$.

Write $a = \lambda b + r$ with either $r = 0$, or $|r| < |b|$, $\lambda \in Z$.

Then $$a/p = \frac{\lambda b}{p^k} + \frac{r}{p^k}.$$

$$\frac{\lambda b}{p^k} = \frac{\lambda}{p^{k-1}} \cdot \frac{b}{p^l} = \frac{\lambda}{p^{k-1}} \cdot q'$$

So $q = tq' + r/p^k$, with $t \in R$.

Now $d\left(\frac{r}{p^k}\right) \leq |r|$ (depending on whether p is a factor in r or not)

and so $d\left(\frac{r}{p^k}\right) \leq |r| < |b| = d(q')$.

Hence R is a Euclidean domain.

2.11 PRINCIPAL IDEAL DOMAINS (PID)

1. Show that the ring $R = \{m/n \mid m, n \in Z, n \text{ odd}\}$ is a principal ideal domain.

Solution: Since R is a subring of Q, R is a domain. Let I be an ideal of R. Every non-zero element of R can be written as $2^k m'/n'$, m', n' odd. Choose an element $a_0 \in I$ with least k, say k_0, i.e. $a_0 = 2^{k_0} u$, u being unit in R. If $a \in I$, $a = 2^k_v$, v unit, $k \geq k_0$ and $a = 2^{k-k_0}\, u^{-1}\, v\, a_0 \in (a_0)$. Thus I is a principal ideal $I = (a_0)$ and

R is a PID.

2. Show that a subring of a principal ideal domain need not be a principal ideal domain.

Solution: We shall show in Problem 4, Section 2.13 that $Q[x]$ is a principal ideal domain but $Z[x]$ is not a PID.

3. Let R be a principal ideal domain. Show that any ideal of a quotient ring of R is principal. If P is a prime ideal, show that R/P is a principal ideal domain.

Solution: Let R/I be a quotient ring and J/I an ideal of R/I, where J is an ideal of R, $J \supset I$. Since R is PID, $J = (a)$ and $J/I = (\bar{a})$, where $\bar{a} = a + I$. If P is a prime ideal, R/P is also a domain.

4. If R is a PID and $p \in R, p \neq 0$, show that the following conditions are equivalent: (*i*) p is prime, (*ii*) p is irreducible, (*iii*) Rp is a prime ideal, (*iv*) Rp is a maximal ideal.

Solution: (*i*) and (*iii*) are equivalent by definition. Also (*ii*) implies (*iv*). Any ideal $I \supset Rp$ is of the form $I = Ra$, $a \mid p$, p is irreducible. Hence a is a unit or an associate of p, *i.e.* $I = R$ or Rp. Hence Rp is maximal. (*i*) $\Rightarrow$ (*ii*) always. (*ii*) $\Rightarrow$ (*iv*) $\Rightarrow$ (*iii*) $\Rightarrow$ (*i*). Thus all are equivalent.

5. Show that in a PID, every non-zero prime ideal is maximal.

Solution: Let P be a non-zero prime ideal in R. The $P = Rp$, $p \neq 0$ prime. Then $P = Rp$ is maximal by the above problem.

6. Let $f : R \to R'$ be a homomorphism of R onto R', which is not an isomorphism. If R is a PID, and R' a doamin, show that R' must be a field.

Solution: Let Ker $f = P$. P is a prime ideal since R' is a domain, and $P \neq 0$, since f is not an isomorphism. Hence P is a maximal ideal of R, *i.e.* $R/P \approx R'$ is a field.

7. Show that every ideal in a PID is contained in a maximal ideal.

Solution: Let I be a proper ideal in a PID, $I = Ra$. a is a non-unit. Let p be an irreducible factor of a, then $Ra \subset Rp$, and Rp is maximal.

8. Show that if d is the g.c.d. of $q_1, ..., q_n$ in a PID, then $d = \lambda_1 q_1 + ... + \lambda_n q_n$, $\lambda_i \in R$.

Solution: Clearly $(q_1) + \ldots + (q_n)$ is an ideal in R and since R is a PID this ideal is equal to Rd, for some $d \in R$.

Then * d is the g.c.d. of

$$q_1, q_2, ..., q_n \text{ and } d = \lambda_1 q + ... + \lambda_n q_n, \lambda_i \in R.$$

9. Let $S \subset R$ be a subring such that both R and S are PID's and let $a, b \in S$. Show that the g.c.d. of a and b in S is the same as the g.c.d. of a and b in R.

Solution: Let d be the g.c.d. of a and b in S. Then $d = \lambda a + \mu b$, $\lambda, \mu \in S \subset R$. Since $d \mid a$, $d \mid b$ in S, we have $d \mid a$, $d \mid b$ in R. Let $d' \mid a$, $d' \mid b$ in R. So that $d' \mid \lambda a + \mu b = d$. Hence d is the g.c.d. of a and b in R.

10. Define a Noetherian ring as a commutative ring in which every increasing sequence of ideals $I_0 \subset ..., I_1 \subset I_n \subset ...$ is stationary, *i.e.* $I_m = I_{m+1} = ...$ Show that a PID is a Noetherian ring.

Solution: Let $I_0 \subset I_1 \subset ... \subset I_n \subset ...$ be an increasing sequence of ideals in a PID. Then $I = I = \cup I_i$ is an ideal and hence equal to Ra, for some $a \in R$. If $a \in I_m$, then

$$Ra \subset I_m \subset I_{m+1} \subset ... \subset I = Ra,$$

i.e. $$I_m = I_{m+1} = ...$$

11. Let R be a PID. Show that there exists a function f on the non-zero elements of R with non-negative integer values such that:

(*i*) $f(a) \le f(b)$ whenever $a \mid b$ and equality holds only when a and b are associates.

(*ii*) For any $a, b \in R$, where neither divides the other, there exist $p, q, r \in R$ such that $r = pa + qb$ with $f(r) < \min(f(a), f(b))$.

Solution:

(*i*) Let $a = \prod_{i=1}^{k} p_i^{\alpha_i}$ and $b = \prod_{i=1}^{k} p_i^{\beta_i}$, $\alpha_i \ge 0, \beta_i \ge 0$

Since $a \mid b, \alpha_i \le \beta_i, \forall\ i$. Define $f(a) = \sum_{i=1}^{k} \alpha_i \ge 0$.

Since $a \mid b$, $\sum \alpha_i \le \sum \beta_i$, *i.e.* $f(a) \le f(b)$.

Suppose further that $f(a) = f(b)$. Then $\Sigma\alpha_i = \Sigma\beta_i$ and since $\alpha_i \le \beta_i$, $\forall\ i$, it follows that $\alpha_i = \beta_i$, *i.e.* $a = b$ (up to unit), *i.e.* a and b are associates.

(*ii*) Let $a, b \in R$ such that neither divides the other so that $(a) \not\subset (a) + (b)$ and $(b) \not\subset (a) + (b)$.

Write $(a) + (b) = (r)$. Then $r = pa + qb$ for some $p, q \in R$. Moreover, since $(a) \not\subset (r)$ and $(b) \not\subset (r)$ r is a proper divisor of a and b and so $f(r) < f(a)$ and $f(r) < f(b)$).

Hence $f(r) < \text{Min}\ (f(a), f(b))$.

12. Show that a commutative ring is Noetherian if, and only if, every ideal I is finitely generated, *i.e.* there exist $a_1, ..., a_n \in I$ such that

$$I = Ra_1 + Ra_2 + ... + Ra_n = (a_1) + (a_2) + ... + (a_n).$$

Solution: Let R be Noetherian and I and ideal of R. Choose $a_1 \in I$, and consider $(a_1) \subset I$. If equality does not hold, choose $a_2 \in I$, $a_2 \notin (a_1)$ and consider $(a_1) + (a_2) \subset I$. Continue the process and after a finite number of steps

$$(a_1) + (a_2) + ... + (a_n) = I,$$

for otherwise there will be a strictly increasing infinite sequence of ideals of R, contradicting the fact that R is Noetherian. Conversely assume that every ideal I is finitely generated and consider a sequence.

$$I_1 \subset I_2 \subset ... \subset I_n \subset ...$$

Let $$I = \bigcup_{1}^{\infty} I_i.$$

Then I is an ideal and hence generated by $a_1, ..., a_n$. Choose k such that

$$a_i \in I_k,\ 1 \le i \le n.\ I \subset I_k \subset I_{k+1} \subset ... I.$$

Then $I_k = I_{k+1} = ...$ Hence R is Noetherian.

13. Which of the following is true? Why?

(*i*) Quotient ring of a PID is a PID.

(*ii*) Product of two PID's is a PID.

Solution:

(*i*) Quotient ring of a PID by a prime ideal is a PID.

(*ii*) Product of two PID's may not be a PID as the product may have zero divisors say in $Z \times Z$.

2.12 UNIQUE FACTORISATION DOMAIN (UFD)

1. Show that in a UFD any two elements a and b have an 1.c.m. $[a, b]$ and $[a, b]$ $(a, b) = ab$.

Solution: Let $$a = p_i^{\alpha_1} \ldots p_k^{\alpha_k} \text{ and } b = p_1^{\beta_1} \ldots p_k^{\beta_k},\ \alpha_i, \beta_i \geq 0.$$

Then $$[a, b] = p_1^{\gamma_1} \ldots p_k^{\gamma_k},\ \gamma_i = \text{Max}\ \{\alpha_i, \beta_i\}$$

and $$(a, b) = p_1^{\delta_1} \ldots p_k^{\delta_k},\ \delta_i = \text{Min}\ \{\alpha_i, \beta_i\}.$$

Then $\gamma_i + \delta_i = \alpha_i + \beta_i$, so that $[a, b]\ (a, b) = a \cdot b$.

2. Show that in a UFD, $a \mid c$, $b \mid c$ and $(a, b) = 1$ implies $ab \mid c$.

Solution: Let $c = p_1^{\alpha_1} \ldots p_k^{\alpha_k}$, $\alpha_i > 0$, p_i irreducible.

Since $a \mid c$, a is a product of some of the factors p_i's with exponents utmost α_i and the same is true of b. Since $(a, b) = 1$, the irreducible factors of a and b cannot have any element in common, so that ab is also a product of some p_i's with exponent utmost α_i. Hence $ab \mid c$.

3. Show that in a UFD, $(a, c) = 1$ and $(b, c) = 1$ implies $(ab, c) = 1$.

Solution: The irreducible factors of c are disjoint with the irreducible factors of a as well as b and hence with the irreducible factors of ab. Then $(a\,b, c) = 1$.

4. If $d = (a, b)$ show that $(a/d, b/d) = 1$.

Solution: Let $$a = \prod_1^k p_i^{\alpha_i} \text{ and } b = \prod_1^k p_i^{\beta_i},\ \alpha_i, \beta_i \geq 0.$$

Then $$d = \Pi\ p_i^{\delta_i},\ \delta_i = \text{Min}\ \{\alpha_i, \beta_i\}.$$

Then either $\alpha_i = \delta_i$ or $\beta_i = \delta_i$ so that either a/d has no factor containing p_i or b/d has no factor containing p_i. This is true for each i. Thus a/d and b/d have no common factors. *i.e.* $(a/d, b/d) = 1$.

5. Show that $[a, (b, c)] = ([a, b], [a, c])$, for any $a, b, c \in R$, a UFD.

Solution: Let $$a = \prod_1^k p_i^{\alpha_i},\ b = \prod_1^k p_i^{\beta_i}$$

and $$c = \prod_1^k p_i^{\gamma_i},\ \alpha_i, \beta_i, \gamma_i \geq 0$$

The relation amounts to showing that

$$\operatorname*{Max}_i \left\{\alpha_i, \operatorname*{Min}_i \{\beta_i, \gamma_i\}\right\} = \operatorname*{Min}_i \left(\operatorname*{Max}_i [\alpha_i, \beta_i], \operatorname*{Max}_i [\alpha_i, \gamma_i]\right)$$

which is true.

6. Let R be a UFD, and $p \in R$, $p \neq 0$, a non-units such that for each $a \in R$ either $p \mid a$ or $(p, a) = 1$. Show that p is irreducible.

Solution: Suppose p is reducible, *i.e.* $p = p_1 p_2$, where p_1 and p_2 are proper factors of p. Apply the condition with $a = p_1$. Now $p \chi a$ as p_2 is a non-unit. Also $(p, a) \neq 1$as p_1 is a non-trivial common factor. This is a contradiction.

7. Let R be a UFD, p an irreducible element and $q = p^n$. Show that if $q \mid ab$, then either $q \mid a$ or $q \mid b^m$ for some m. Deduce that the ideal generated by q is a primary ideal.

Solution: Let $\quad a = \prod_1^k p_i^{\alpha_i}$ and $b = \prod_1^k p_i^{\beta_i}$, $\alpha_i, \beta_i \geq 0$

Since $q \mid ab$, $p \mid ab$, p must be equal to one of the p_i's say $p = p_1$. If $\alpha_1 \geq n$, then $q \mid a$. Otherwise $\beta_1 > 0$.

Hence for sufficiently large m, $q \mid b^m$. This implies that (q) is also primary.

8. If in a factorisation domain any two elements a and b have a.g.c.d., which can be expressed as $d = \lambda a + \mu b$, show that R is a UFD.

Solution: It is sufficient to show that every irreducible element is prime. Let p be irreducible and $p \mid ab$. Let $(p, a) = d$. If $d = p$, then $p \mid a$. If this is not the case, $d = 1$ as p is irreducible.

By assumption $\quad 1 = \lambda p + \mu a \quad$ for some $\quad \lambda, \mu \in R$.

So $\quad b = \lambda p d + \mu a b$.

Since $p \mid ab$, $p \mid b$. Hence p is a prime.

9. Let R be a principal ideal domain and $I \neq 0$ and ideal in R. Show that I can be expressed uniquely as $I = P_1^{\alpha_1} P_2^{\alpha_2} \ldots P_n^{\alpha_n}$, where P_i's are prime ideals. (Such rings are called Dedekind domains.)

Solution: Since R is a PID, $I = (a)$, $a \neq 0$. Write $a = p_i^{\alpha_1} \cdots p_n^{\alpha_n}$ uniquely, p_i irreducible.

Let $\quad P_i = (p_i)$. Then each P_i is a prime ideal and

$$I = P_1^{\alpha_1} \ldots P_n^{\alpha_n}, \text{ uniquely.}$$

10. Let R be a UFD and K the quotient field of R. Show that every $\alpha \in K$ which satisfies a relation of the type

$$\alpha^n + a_1 \alpha^{n-1} + a_2 \alpha^{n-2} + \ldots + a_n = 0,$$

$a_i \in R$ must lie in R. (R is integrally closed in K.)

Solution: Write $\alpha = p/q$, $p, q \in R$, $q \neq 0$, $(p, q) = 1$.

Then $\quad (p/q)^n + a_1 (p/q)^{n-1} + \ldots + a_n = 0$

i.e. $\quad p^n + a_1 p^{n-1} q + \ldots + a_n q^n = 0$

We claim that q is a unit for if it has an irreducible factor p_1, then p_1 divides also

$$p^n = -(a_1p^{n-1}q + ... + a_n q^n)$$

and hence $p_1 \mid p$. This contradicts the assumption that $(p, q) = 1$. Hence q is a unit in R and $\alpha = p/q \in R$.

11. Show that every Noetherian ring R, which is an integral domain is a factorisation domain?

Solution: Let $a \in R, a \neq 0$. If a is irreducible there is nothing to prove. Otherwise $a = b \cdot c$, where b and c are proper divisors of a. If both b and c are irreducible, then $a = b \cdot c$ is the required factorisation. If not, b (or c) is reducible and write $b = e \cdot f$, where e, f are proper divisors. If we continue this process, after a finite number of steps all factors will be irreducible for otherwise there will be an infinite sequence of elements $a = a_0, a_1, ... a_n, ...$ such that each a_{n+1} is a proper divisor of a_n, *i.e.* there will be an infinite sequence of ideals

$$(a_0) \subsetneq (a_1) \subsetneq ... \subsetneq (a_n) \subsetneq ...$$

a contradiction, since R is Notherian.

12. Give an example to show that in a UFD, the g.c.d. of two elements a and b need not be expressible in the form $\lambda a + \mu b$, $\lambda, \mu \in R$.

Solution: Consider the ring $R = Z[x]$. Then $Z[x]$ is a UFD. Let $a = 2$ and $b = x$. Then $(a, b) = 1$. But 1 cannot be expressed as a combination of 2 and x for, if $1 = 2f + g.x$, then $g = 0$. Then f is constant polynomial, which is a contradiction as f belongs to Z.

2.13 POLYNOMIAL RINGS

1. Find the g.c.d. of

$$f(x) = x^3 - 2x^2 + 6x - 5 \in Q[x]$$

and

$$g(x) = x^2 - 2x + 1 \in Q[x].$$

Solution: $x^3 - 2x^2 + 6x - 5 = x(x^2 - 2x + 1) + 5x - 5$

$$x^2 - 2x + 1 = (5x - 5)\frac{1}{5}(x - 1)$$

Hence, the g.c.d. is $x - 1$ (up to units).

2. What are the units in $Z_7[x]$?

Solution: The units is $Z_7[x]$ are the units of Z_7 *i.e.* Z_7^*, the non-zero elements of Z_7.

3. Show that for any R, $R[x]$ can never be a field.

Solution: If $R[x]$ is a field, x have an inverse *i.e.* $x f(x) = 1, f(x) \in R[x]$. This is impossible if R is a domain, as $\deg(x) \geq 1$. If R has zero divisors, $R[x]$ also has zero divisors *i.e.* it cannot be a field.

4. Why is $Z[x]$ not a Euclidean domain? Where does the proof of the fact that $F[x]$ (F field) is a Euclidean domain fail in this case?

Solution: $Z[x]$ has ideals which are not principal say $I = (2) + (x)$ and hence cannot be a Euclidean domain. The division algorithm proof fails when the dividing polynomial has highest degree coefficient, a non-unit in Z.

5. Given examples of two polynomials $f(x)$ and $g(x)$ such that $\deg(fg) < \deg(f) + \deg(g)$.

Solution: Let $$R = Z_4, f(x) = 2x = g(x).$$
$$f(x)\, g(x) = 0 \text{ has deg } 0.$$

6. Le R and R' are isomorphic rings, show that $R[x]$ and $R'[x]$ are also isomorphic.

Solution: Let $\theta : R \to R'$ be an isomorphism

where $$\theta(a) = \bar{a},\ a \in R,\ \bar{a} \in R'.$$

Define the map $\bar{\theta} : R[x] \to R'[x]$ given by $\bar{\theta}\left(\sum_i a_i x^i\right) = \sum_i \bar{a}_i x^i$.

Then $\bar{\theta}$ is a ring homomorphism which is also an isomorphism.

7. Let F be a field and $I = \{f(x) \mid f(x) \in F[x], f \text{ has no constant term}\} \cup \{0\}$. Show that I is an ideal in $F[x]$. Is it a maximal ideal?

Solution: Clearly I is closed for the operations of – and multiplication by any element of $F[x]$. Hence, it is an ideal. It is a maximal ideal for if $J \supsetneq I$, choose $f \in J, f \notin I$

$$f = a_0 + \sum_{i>0} a_i x^i,\ a_0 \neq 0 \quad \text{and} \quad f - \sum_{i>0} a_i x^i = a_i x^i = a_0 \in J,$$

a_0 a unit. Hence $J = F[x]$ and I is maximal.

8. Let $F = Q$ and $I = \{f(x) \mid f(3) = 0\}$. Show that I is an ideal in $F[x]$. Write it as a principal ideal.

Solution: Clearly I is closed for the operations of – and multiplication by any element of $F[x]$. If $f(x) \in I, f(3) = 0$ and hence $f(x)$ is divisible by $x - 3$. Hence I is generated by $x - 3$.

9. Let $F = Q$ and $I = \{f(x) \mid f(4) \neq 0\}$. Is I an ideal in $F[x]$? Why?

Solution: No. $f(x) = x - 2$ and $g(x) = x^2 - x - 10$ belong to I but
$$g(x) - f(x) = x^2 - 2x - 8 \notin I.$$

10. Let F be a field. Show that every non-zero prime ideal of $F[x]$ is maximal.

Solution: Since $F[x]$ is a PID, any non-zero prime ideal of $F[x]$ is maximal.

11. Let $f(x) = a_0 + a_1 x + \ldots + a_n x^n$ and
$f'(x) = a_1 + 2a_2 x + \ldots + n a_n x^{n-1}$
be in $Q[x]$. Show that f is divisible by the square of a polynomial if, and only if, $f(x)$ and $f'(x)$ are not relatively prime.

Solution: It is easy to verify that
$$(f + g)' = f' + g' \text{ and } (fg)' = fg' + f'g.$$

Suppose f is divisible by the square of a polynomial $d(x)$.

Then $$f(x) = d(x)^2 g(x).$$

Then $$f'(x) = 2d(x) d'(x) g(x) + d(x)^2 g'(x).$$

Therefore $f(x)$ and $f'(x)$ have a non-trivial common factor $d(x)$. Conversely assume that $f(x)$ and $f'(x)$ have a non-trivial common factor $d(x)$. Then they have an irreducible factor in common, say $p(x)$. Write

$$f(x) = p(x)^\lambda g(x), \lambda \geq 1, (p(x), g(x)) = 1$$

Then $$f'(x) = \lambda p(x)^{\lambda-1} p'(x) g(x) + p(x)^\lambda g'(x).$$

Since $p(x) \mid f'(x)$ and $\lambda \geq 1$, $p(x) \mid \lambda p(x)^{\lambda-1} p'(x) g(x)$

Since $$\deg p' < \deg p, (p, p') = 1 \text{ and } (p, g) = 1.$$

So, $p(x) \mid \lambda p(x)^{\lambda-1}$ and hence $\lambda \geq 2$.

Hence $f(x)$ has a square factor $p(x)^2$.

12. Let $R = Q[x]$ and $I = R(x^2 + 2)$, the ideal of all multiples of $x^2 + 2$. Is R/I an integral domain? Why?

Solution: The polynomial $x^2 + 2 \in Q[x]$ is irreducible and hence I is a prime ideal. Thus R/I is an integral domain.

13. Let F be the field of real numbers and $I = (x^2 + 1) F[x]$. Show that I is a maximal ideal and $F[x]/I$ is isomorphic to the field of complex numbers.

Solution: $x^2 + 1 \in F[x]$ is irreducible and hence I is a maximal ideal. Thus $F[x]/I$ is a field. Consider the map $\theta : F[x] \to C$ given by $\theta(x) = i \in C$ and extended by $\theta(\Sigma a_k x^k) = \Sigma a_k (i)^k$. Then θ is a ring homomorphism, "onto" with Ker $\theta = I$. Thus $F[x]/I \approx C$.

14. Let $f(x), g(x) \in F(x)$ and I defined by

$$I = \{h(x) \mid h(x) = p(x) f(x) + q(x) g(x), p, q \in F[x]\}$$

Show that I is an ideal in $F[x]$. If I is generated by $d(x)$, how is $d(x)$ related to $f(x)$ and $g(x)$?

Solution: I is closed for the operation of – and multiplication by an element of $F[x]$. Hence I is an ideal. If $I = (d(x))$ then $d(x)$ is the g.c.d. of $f(x)$ and $g(x)$.

15. Let R be a commutative ring with unit and $R' = R[x]$. Then the ring $R'[y]$ is called the polynomial ring in two variables x and y over R and is denoted by $R[x, y]$. Show that

(*i*) $R[x, y] = R''[x]$ where $R'' = R[y]$.

(*ii*) $R[x, y]$ is also commutative with unit element.

(*iii*) If R is a domain, $R[x, y]$ is also domain.

(*iv*) If F is a field, $F[x, y]$ is not a Euclidean domain by showing that the ideal

$$I = \{x f(x, y) + y g(x, y) \mid f, g \in F[x, y]\}$$

is not a principal ideal.

Solution:

(*i*) If $f \in R[x, y], \quad f = \sum_i a_i(x)\, y^i$

where $a_i(x) = \sum_j p_{ij} x^j$, so that

$$f = \sum_i \left(\sum_j p_{ij} x^j\right) y^i = \sum_j \left(\sum_i p_{ij} y^i\right) x^j$$

$$= \sum_j b_j(y)\, x^j \in R''[x],$$

where $b_j(y) = \sum_i b_{ij} y^i \in R[y] = R''$

(*ii*) Let $f \in R[x, y]$

$$f(x, y) = \sum_{i,j} a_{ij} x^i y^j \text{ and } g(x, y) = \sum_{i,j} b_{ij} x^i y^j.$$

Since $a_{ij} b_{kl} = b_{kl} a_{ij}$ for all i, j, k, and l.

$f(x, y)\, g(x, y) = g(x, y)\, f(x, y)$. Thus $R[x, y]$ is commutative. If 1 is the unit element of R, it is also the unit element of $R[x, y]$.

(*iii*) If R is a domain so is $R[x]$. Hence $R[x][y] = R[x, y]$ is also domain.

(*iv*) Suppose I is a principal ideal generated by $\theta(x, y)$. Since $x \in I$ ($f = 1, g = 0$). $\theta(x, y)$ must be of degree utmost one in x. Similarly, since $y \in I$, $\theta(x, y)$ must have degree utmost one in y. Hence $\theta(x, y) = a + bx + cy$.

Since $x \in I$, $c = 0$ and since $y \in I$, $b = 0$ *i.e.*, q (x, y) is a unit, a contradiction.

16. Define the polynomial ring $R[x_1, x_2, \ldots, x_n]$ and prove properties (*ii*) and (*iii*) of Exercise 15 for this ring.

Solution: $R[x_1, \ldots, x_n] = R'[x_n]$ where $R' = R[x_1, \ldots, x_{n-1}]$ is defined by induction on n.

(*ii*) If R is commutative, so is $R[x_1]$. By induction $R[x_1, \ldots, x_n]$ is commutative.

(*iii*) If R is a domain, $R[x_1]$ is a domain. By induction $R[x_1, \ldots, x_n]$ is a domain.

17. Let I be an ideal of a commutative ring R and $I[x, y]$ those polynomials in x, y with coefficiets in I. Show that $I[x, y]$ is an ideal of $R[x, y]$ and that

$$\frac{R[x, y]}{I[x, y]} \approx \frac{R}{I}[x, y].$$

Solution: $I[x, y]$ can be verified to be an ideal using the fact that I is an ideal in R. Consider the projection map $R \to R/I$. This induces a ring homomorphism $R[x, y] \to R/I\,[x, y]$ which is "onto" and has kernel $I[x, y]$. Hence

$$\frac{R[x, y]}{I[x, y]} \approx \frac{R}{I}[x, y].$$

2.14 ROOTS OF POLYNOMIALS

1. Show that every polynomial of odd degree with real coefficients has at least one real root.

Solution: Since the complex roots of a polynomial with real coefficients occur in conjugate paris there must be a real root.

2. Show that any irreducible polynomial over reals is of degree utmost 2.

Solution: The polynomial must be of even degree, otherwise it will have a real root and hence reducible unless it is linear. Moreover, since complex roots occur in conjugate pairs, it will have quadratic real polynomials as factors and there must be only one such factor as the polynomial is irreducible.

3. Let $f(x) \in F[x]$. Show that for any $\alpha \in F$, $(x - \alpha)$ divides $f(x) - f(\alpha)$.

Solution: Since α is a root of $g(x) = f(x) - f(\alpha)$, $(x - \alpha)$ must divide $g(x)$.

4. Factorise $4x^2 - 4x + 8$ as a product of irreducible elements in $Q[x]$.

Solution: $4x^2 - 4x + 8 = 4(x^2 - x + 2)$

Now $x^2 - x + 2$ is irreducible over Q as it has complex roots.

5. Find all irreducible polynomials of degree less than or equal to 3 in $Z_2[x]$.

Solution: Linear : x and $x + 1$

Quadratic : $x^2 + x + 1$

Cubic : $x^3 + x^2 + 1$ and $x^3 + x + 1$

6. Find all the roots of $f(x) = x^3 + x^2 + x + 1$ in Z_5.

Solution: The roots in Z_5 are $\overline{2}, \overline{3}$ and $\overline{4}$.

7. Find whether the following pairs of polynomials have a common root in Q.

(*i*) $x^3 + x - 2$; $x^4 - x^3 - 2x + 2$.

(*ii*) $x^2 + x + 2$; $x^5 - x^4 + x^3 - x^2 + x - 1$.

Solution: (*i*) 1 is a common root of both the polynomials.

Also $x^3 + x - 2 = (x - 1)(x^2 + x + 2)$ and $x^2 + x + 2$ has no other rational root

(*ii*) $x^2 + x + 2$ has no rational root and hence there is no common root.

8. If α is a root of

$$f(x) = a_0 + a_1 x + \ldots + a_n x^n \in F[x],$$

show that $1/\alpha$ is a root of

$$g(x) = a_n + a_{n-1} x + \ldots + a_0 x^n$$

Solution:

$$g(1/\alpha) = a_n + \frac{a_{n-1}}{\alpha} + \ldots \frac{a_0}{\alpha^n}$$

$$= \frac{1}{\alpha^n}(a_0 + a_1 \alpha + \ldots + a_n \alpha^n)$$

$$= \frac{f(\alpha)}{\alpha^n} = 0.$$

9. Let $f(x) \in Z[x]$ be a polynomial with highest degree coefficient 1. Show that if $f(x)$ has a root in Q, it must be an integer.

Solution: Let $f(x) = a_0 + a_1 x + \ldots + a_{n-1} x^n + x^n$; $a_i \in Z$.

Let $\alpha = p/q$ be a root of $f(x)$ with $(p, q) = 1$.

Then $a_0 + a_1 p/q + \ldots + p^n/q^n = 0$

So $a_0 q^n + a_1 q^{n-1} p + \dots + p^n = 0$

If q is not a unit, it has an irreducible factor q_1 and then q_1 divides $-(a_0 q^n + \dots + a_{n-1} q) = p^n$. Hence q_1 divides p^n, *i.e.*, $q_1 \mid p$. This contradicts the assumption that $(p, q) = 1$. Hence q is a unit, and $\alpha = p \mid q \in Z$.

10. Let F be a finite field. Construct a polynomial over F which has none of the element of F as root.

Solution: Let $a_1, a_2, \dots, a_n$ be the elements of F. The required polynomial is

$$f(x) = 1 + \prod_{i=1}^{n} (x - a_i).$$

2.15 FACTORISATION OF POLYNOMIALS

1. Let $f(x) = x^2 + 8x - 2$. Show that it is irreducible over Q. Is it irreducible over reals? Why?

Solution: $f(x)$ has no rational root. Hence it is irreducible over Q. It is irreducible over reals because $f(x)$ has real roots.

2. Test for irreducibility over Q the following polynomials.

(*i*) $8x^3 + 6x^2 - 9x + 24$.

(*ii*) $x^4 + 9x + 3$.

Solution: (*i*) Yes. Apply Eisenstein's criterian with $p = 3$.

(*ii*) Yes. Apply Eisenstein's criterian with $p = 3$.

3. Let R be a UFD and $f(x), g(x) \in R[x]$ such that $f(x)$ is primitive. Show that if $f(x) \mid ag(x), a \in R$ then $f(x) \mid g(x)$.

Solution: Let $p(x)$ be an irreducible factor of $f(x)$. Since $f(x)$ is primitive $p(x)$ is also primitive. Also $p(x) \mid ag(x)$. Now $(p(x), a) = 1$, since $p(x)$ is primitive. Hence $p(x) \mid g(x)$. Hence every irreducible factor of $f(x)$ divides $g(x)$. Since $R[x]$ is a UFD and $f(x)$ divides a $g(x)$, the exponent of $p(x)$ in the factorisation of $f(x)$ is utmost the exponent of $p(x)$ in the factorisation of $g(x)$. Hence $f(x) \mid g(x)$.

4. Factorise $x^3 + 1$ as a product of irreducible elements of $Z_7[x]$.

Solution: $x^3 + 1$ has $\overline{3}$ as a root. Hence $x - 3 \mid x^3 + 1$ over $Z_7[x]$.

Write $x^3 + 1 = (x - 3)(x^2 + 3x + 2)$.

Now $x - 5$ divides $x^2 + 3x + 2$ as $x^2 + 3x + 2 = (x - 5)(x + 1)$.

Thus $x^3 + 1 = (x - 3)(x - 5)(x + 1)$ in $Z_7[x]$.

5. Express $f(x) = 2x^2 - 8x + 6$ as a product of its content and a primitive polynomial in $Z[x]$.

Solution: The content of $f(x)$ is 2 and so

$$f(x) = 2(x^2 - 4x + 3) \text{ and } x^2 - 4x + 3$$

is a primitive polynomial.

6. (*i*) If R is a principal ideal domain, Show that $R[x]$ in general is not a PID.

(*ii*) Show that if $R[x]$ is a PID, then R must be a field.

Solution: (*i*) If $R = Z$, $Z[x]$ is not a PID as the ideal of all polynomials with even constant term is not principal, because $I = (2, x) = (\theta(x))$, say. Now, both 2 and x cannot be a multiple of $\theta(x)$.

(*ii*) Suppose $R[x]$ is a PID. Consider the map $\theta : R[x] \to R$ defined by $\theta(\Sigma a_i x^i) = a_0$. θ is a ring homomorphism of $R[x]$ onto R, which is not an isomorphism. So the kernel P of θ is a non-zero prime ideal, which is maximal.

Thus $\dfrac{R[x]}{P} \approx R$ is a field.

7. Is $f(x) = x^4 - 2$ irreducible over the ring of Gaussian integers? Why?

Solution: The roots of $f[x]$ in C are $\pm\sqrt{\sqrt{2}}$ and $\pm i\sqrt{\sqrt{2}}$. None of the products of the linear factors $x \pm \sqrt{\sqrt{2}}$, $x \pm i\sqrt{\sqrt{2}}$ have coefficients in the ring of Gaussian integers. Hence the polynomial is irreducible.

8. Factorize $x^p - x$ into irreducible polynomials in $Z_p[x]$.

Solution: $x^p - x = x \prod_{i=1}^{p-1} (x - \bar{i})$ is the irreducible factorisation.

9. Show that the field $Z_p(x)$ of rational functions in x over Z_p is an infinite field of ch p.

Solution: Since ch $Z_p = p$, ch $Z_p(x)$ is also p. Moreover, it is an infinite field as there are polynomials of arbitrarily large degree.

10. For any integral domain R, show that there are infinitely many irreducible polynomials over R with highest degree coefficient 1.

Solution: Suppose there are only finitely many irreducible polynomials $p_1, p_2, \ldots, p_k$ of degrees $n_1 \ldots n_k$.

Then $q(x) = 1 + \prod_{i=1}^{k} p_i \in R[x]$ is either irreducible, or has an irreducible factor.

None of p_i's are factors of q, nor q can be any of the p_i's as its degree is greater than the degree of any p_i. This is a contradiction.

11. Let $f(x) = a_0 + a_1 x + \ldots + a_n x^n \in R[x]$ be a polynomial of degree n and p an irreducible element such that for some $k < n$,

$$p \mid a_0, p \mid a_1, \ldots p \mid a_{k-1}, p \chi a_k a_n, p^2 \chi a_0.$$

Show that $f(x)$ has at least one irreducible factor of degree $\geq k$.

Solution: If $f(x)$ is irreducible f is a factor of f. Else write $f = gh$. Let $f(x)$

$$= \sum_0^n a_i x^i, g(x)$$

$$= \sum_0^r b_i x^i, h(x) = \sum_0^s c_i x^i.$$

Then $$a_i = b_i c_0 + b_{i-1} c_1 + \ldots + b_0 c_i$$

$a_0 = b_0 c_0$. Since $p \mid a_0$, $\mid b_0$ or $p \mid c_0$ but not both as $p^2 \nmid a_0$. Assume $p \mid b_0$ and $p \nmid c_0$. Since $p \mid b_0$, p divides the constant term of an irreducible factor of $g(x)$ and so assume $g(x)$ is irreducible. Since $p \nmid a_n$, $p \nmid b_i$ for some i. Choose least such i, *i.e.*, $p \nmid b_i$, but $p \mid b_{i-1}, \ldots p \mid b_0$. We claim that $i \geq k$. For this consider $a_i = b_i c_0 + b_{i-1} c_1 + \ldots + b_0 c_i$.

Since $p \mid b_{i-1}, \ldots p \mid b_0$, and $p \nmid b_i c_0$, it follows that $p \nmid a_0$. Hence by assumption $i \geq k$. In particular $b_i \neq 0$, $(i \geq k)$. Thus the factor $g(x)$ of $f(x)$ has degree $\geq k$.

12. Show that the ideal $I = (3, x^3 - x^2 - 2x - 1)$ in $Z[x]$ is not principal. Is $\dfrac{Z[x]}{I}$ a domain? Why?

Solution: Suppose I is principal, *i.e.*, $I = (f(x))$. Then $3 \in I$ implies that $f(x)$ must divide 3, *i.e.*, $f(x) = \pm 1, \pm 3$. If $f(x) = \pm 1$, $I = Z[x]$ a contradiction, and if $f(x) = \pm 3$, $I = 3Z[x]$. In this case the second polynomial $x^3 - x^2 - 2x - 1$ cannot belong to I. Hence I is not principal. $Z[x]/I$ is a domain as the ideal I is a prime ideal.

CHAPTER 3

Vector Spaces and Modules

3.1 VECTOR SPACES

1. Show that in a vector space V over F

$$(a - b)\, v = av - bv, \quad a, b \in F,\ v \in V$$

$$a\,(v - w) = av - aw, \quad a \in F,\ v, w \in V.$$

Solution: $(a + (-a))\, v = av + (-a) \cdot v$

$$= 0 \cdot v = 0$$

So $(-a)\, v = -(a \cdot v)$

By distributivity $(a - b)\, v = av + (-b)\, v = av - bv$

$$a\,(v + (-v)) = av + a \cdot (-v)$$

$$= a \cdot 0 = 0$$

So $a \cdot (-v) = -(av)$

By distributivity $a\,(v - w) = a \cdot v + a \cdot (-w) = av - aw$

2. Let V be the abelian group of positive real numbers for multiplication. Define scalar multiplication in V by $ax = x^a$, $a \in \mathbb{R}$, $x \in V$. Show that V is a vector space over $\mathbb{R}$.

Solution: $(a + b)\, x = x^{a+b} = x^a \cdot x^b = (ax) \cdot (bx)$

$$a\,(xy) = (xy)^a = x^a \cdot y^a = (ax)\,(ay)$$

$$1 \cdot x = x^1 = x$$

$$(ab)\, x = x^{ab} = (x^b)^a = a\,(bx)$$

3. Can every abelian group V be made into a vector space by the scalar multiplication $av = 0$, $a \in F$, $v \in V$? Why?

Solution: No. All the axioms of the vector space are satisfied except the axiom $1 \cdot v = v$ because $1 \cdot v = 0$, in this case.

4. Let G be an abelian group such that $px = 0$ for all $x \in G$, where p is a fixed prime. Show that G is a vector space over Z_p.

Solution: For $\bar{a} \in Z_p, x \in G$, define $\bar{a} \cdot x = ax$.

This is well defined for if $\bar{a} = \bar{b}$, $p \mid b - a$

and $\quad 0 = px = (b - a)\, x = bx - ax.$

Hence $ax = bx$. All the axioms of the vector space are satisfied.

5. Give an example of a vector space with 125 elements.

Solution: Let $F = Z_5$

and $V = \{(a, b, c) \mid a, b, c \in F\}$ with co-ordinate wise addition and scalar multiplication.

Then V is a vector space over F with 125 elements.

6. Show that the following are vector spaces over reals:

(*i*) The set of all real valued differentiable functions on [0, 1] for usual addition and scalar multiplication.

(*ii*) The set of the all sequences of real numbers

$$x = (x_1, x_2, \ldots, x_n, \ldots) \text{ with } \sum_{i=1}^{\infty} x_i^2 < \infty$$

for co-ordinatewise addition and scalar multiplication.

Solution: (*i*) $(f + g)' = f' + g'$ and $(af)' = af'$ so that the space is closed for the operations. All the axioms of the vector space are satisfied.

(*ii*) By Schwartz inequality

$$\sum_{i=1}^{n} x_i y_i \le \left(\sum_{i}^{n} x_i^2\right)\left(\sum_{1}^{n} y_i^2\right)$$

$$\le \left(\sum_{1}^{\infty} x_i^2\right)\left(\sum_{1}^{\infty} y_i^2\right) < \infty$$

As $n \to \infty \quad \sum_{i=1}^{\infty} x_i y_i < \infty$

Now $\quad \sum_{1}^{\infty} (x_i + y_i)^2 = \sum_{1}^{\infty} x_i^2 + 2\sum_{1}^{\infty} x_i y_i + \sum_{1}^{\infty} y_i^2 < \infty$

and $\quad \sum_{1}^{\infty} (a\,x_i)^2 = a^2 \sum_{1}^{\infty} x_i^2.$

Thus the space is closed for the operations of + and scalar multiplication. All the axioms of the vector space are satisfied.

7. Show that the set of all functions which satisfy the differential equation $\dfrac{d^2 f}{dx^2} + \dfrac{3df}{dx} = 0$ is a vector space.

Solution: If f, g are solutions, so are $f + g$ and af because

$$\frac{d^2}{dx^2}(f + g) + 3\frac{d}{dx}(f + g) = \frac{d^2 f}{dx^2} + \frac{3\,df}{dx} + \frac{d^2 g}{dx^2} + \frac{3\,dg}{dx} = 0$$

$$\frac{d^2}{dx^2}(af) + \frac{3d}{dx}(af) = a\left(\frac{d^2 f}{dx^2} + \frac{3df}{dx}\right) = 0.$$

All the axioms of vector space are satisfied.

8. Let V be a vector space over F and F' a subfield of F. Show that V is also a vector space over F'.

Solution: $a \in F', x \in V$ implies $ax \in V$.

All the axioms of vector space are satisfied.

9. Let V be a vector space over $\mathbb{R}$ and $W = \{(v, v') \mid v, v' \in V\}$. Define addition in W co-ordinatewise and multiplication by complex number by

$$(a + ib)(v, v') = (av - bv', av' + bv).$$

Show that W is vector space over C (complexification of V.)

Solution: W is clearly an abelian group for co-ordinatewise addition. Since V is a vector space over $\mathbb{R}$, using these axioms, we can verify the axioms of vector space W over C.

10. Let X be any set and F any field. Show that the set V of all functions from X to F, is a vector space over F, for the operations

$$(f+g)(x) = f(x) + g(x), (af)(x) = a \cdot f(x), \quad f, g \in V, \quad x \in X, a \in F.$$

Solution: Clearly V is an abelian group for addition with constant zero function as zero element. All the axioms for the vector space are satisfied.

3.2 SUBSPACES

1. If V_1 is a subspace of V_2 and V_2 is a subspace of V, show that V_1 is a subspace of V.

Solution: Subgroup of a subgroup is a subgroup. So V_1 is an abelian subgroup of V. Also for $v \in V_1$ and $a \in F$, $av \in V_2 \subset V$.

2. If V_1 and V_2 are subspaces of V, show that $V_1 \cup V_2$ is not a subspace in general. Show further that $V_1 \cup V_2$ is a subspace if and only if either $V_1 \subset V_2$ or $V_2 \subset V_1$.

Solution: Let $\quad V = \mathbb{R}^2,$

$$V_1 = \{(x, 0) \mid x \in \mathbb{R}\} \quad \text{and} \quad V_2 = \{(0, y) \mid y \in \mathbb{R}\}.$$

V_1, V_2 are subspaces but $V_1 \cup V_2$ is not a subspace. If either $V_1 \subset V_2$ or $V_2 \subset V_1$, $V_1 \cup V_2 = V_1$ or V_2. Hence it is a subspace. Conversely assume $V_1 \cup V_2 = V$ is a subspace. If $V_1 \not\subset V_2$ and $V_2 \not\subset V_1$ there exists $v \in V_1$, $v \notin V_2$ and $w \in V_2$, $w \notin V_1$.

Now $\quad v + w \in V \quad$ but $\quad v + w \notin V_1 \cup V_2$, a contradiction.

3. Let V_1 and V_2 be subspaces of V. Show that $V_1 + V_2$ is the subspace generated by $V_1 \cup V_2$.

Solution: $\quad V_1 + V_2 \supset V_1 \quad$ and $\quad V_1 + V_2 \supset V_2$

and so $\quad V_1 + V_2 \supset V_1 \cup V_2.$

Conversely if W is a subspace containing $V_1 \cup V_2$, $W \supset V_1$ and $W \supset V_2$ and so $W \supset V_1 + V_2$. So $V_1 + V_2$ is the smallest subspace containing $V_1 \cup V_2$ and hence is generated by $V_1 \cup V_2$.

4. (*a*) Which of the following spaces are subspaces of $\mathbb{R}^n$?

(*i*) $\{x = (x_1, x_2, ..., x_n) \mid x_1 > x_2\}$.

(*ii*) $\{x = (x_1, x_2, ..., x_n) \mid x_1 + x_2 + ... + x_{n=0}\}$.

(*iii*) $\{x = (x_1, x_2, ..., x_n) \mid x_1 + x_2 + ... + x_n = 3\}$.

(*iv*) $\{x = (x_1, x_2 ..., x_n) \mid x_1 = 0 \text{ and } x_n = 0\}$.

(*v*) $\{x = (x_1, x_2, ..., x_n) \mid x_1 = 2x_2 + 3x_3 + ... + nx_n\}$.

Solution: (*i*) No (*ii*) Yes (*iii*) No (*iv*) Yes (*v*) Yes.

(*b*) Let $V = C[0, 1]$ be the space of all continuous functions on [0, 1]. Which of the following spaces are subspaces of V?

(*i*) $\{x \mid x \in V, \ x(t) \geq 0\}$.

(*ii*) $\{x \mid x \in V, \ x(t^2) = x(t)^2\}$.

(*iii*) $\{x \mid x \in V, \ x(t) = x(-t)\}$.

(*iv*) $\{x \mid x \in V, x \text{ is a polynomial of degree } 3\}$.

Solution: (*i*) No (*ii*) No (*iii*) Yes (*iv*) No.

5. If V_1, V_2 and V_3 are subspaces of V, show that

$$V_3 \cap (V_1 + V_2) \supset (V_3 \cap V_1) + (V_3 \cap V_2)$$

Give example to show that the above spaces may not be equal.

Solution: $V_1 \subset V_1 + V_2$ and so $V_1 \cap V_3 \subset (V_1 + V_2) \cap V_3$

$V_2 \subset V_1 + V_2$ and so $V_2 \cap V_3 \subset (V_1 + V_2) \cap V_3$

Hence $V_1 \cap V_3 + V_2 \cap V_3 \subset (V_1 + V_2) \cap V_3$.

Example: Equality does not hold when $V = \mathbb{R}^2$

$$V_1 = \{(x, 0) \mid x \in \mathbb{R}\}, \quad V_2 = \{(0, y) \mid y \in \mathbb{R}\}$$

and $V_3 = \{(x, x) \mid x \in \mathbb{R}\}$.

In this case the R.H.S. is {0} and L.H.S. is V_3.

6. What is the subspace generated by (1, 0, 0) and (0, 2, 0) in $\mathbb{R}^3$?

Solution: The x-y plane $z = 0$.

7. Find conditions under which $\{x_1, x_2, x_3\}$ lies in the subspace generated by (2, 3, 1) and (0, 1, 2) in $\mathbb{R}^3$.

Solution: The condition is $\begin{vmatrix} x_1 & x_2 & x_3 \\ 2 & 3 & 1 \\ 0 & 1 & 2 \end{vmatrix} = 0,$

i.e. $5x_1 - 4x_2 + 2x_3 = 0$

8. Let $F = Z_5$. What is the subspace generated by (1, 1, 1) and (2, 2, 0) in IF^3?

Solution: $\begin{vmatrix} x & y & z \\ 1 & 1 & 1 \\ 2 & 2 & 0 \end{vmatrix} = 0,$ *i.e.* $-2x + 2y = 0,$ *i.e.* $x - y = 0$

9. Let S and S' be subsets of V. Show that

(*i*) $S \subset S'$ implies $L(S) \subset L(S')$

(*ii*) $L(S \cup S') = L(S) + L(S')$

Solution: (*i*) Clearly $S \subset S'$ implies $L(S) \subset L(S')$

(*ii*) Since $S \subset S \cup S'$, $L(S) \subset L(S \cup S')$

and $S' \subset S \cup S'$ implies $L(S') \subset L(S \cup S')$.

So $L(S) + L(S') \subset L(S \cup S')$.

Conversely let $v \in L(S \cup S')$

and write $$v = \sum_{i=1}^{n} a_i v_i, \; v_i \in S \cup S'.$$

$$= \sum_{v_i \in S} a_i v_i + \sum_{w_i \in S'} a_i w_i \in L(S) + L(S').$$

So $$L(S \cup S') = L(S) + L(S').$$

10. Let V be a vector space over F such that the only subspaces of V and $\{0\}$ and V. What can be say about the nature of V?

Solution: V must consist of scalar multiples of one single non-zero vector.

11. Find all subspaces of $V = \mathbb{R}^3$.

Solution: Subspaces of $\mathbb{R}^3$ are

(*i*) $\mathbb{R}^3$

(*ii*) Planes defined by a single linear equation

(*iii*) Lines defined by 2 independent linear equations and

(*iv*) The origin (0, 0, 0).

3.3 LINEAR DEPENDENCE, INDEPENDENCE AND BASIS

1. Determine whether the following elements of $V = \mathbb{R}^3$ are linearly independent.

(*i*) (1, 0, 1), (1, 1, 1) and (0, 0, 1)

(*ii*) (2, – 1, 3), (4, 1, – 1) and (2, 3, – 3).

Solution: (*i*) Linearly independent as any linear combination of the first and the first and the third vector has middle co-ordinate zero and (1, 1, 1) does not lie in this space

(*ii*) The determinant of the row vectors is

$$\begin{vmatrix} 2 & -1 & 3 \\ 4 & 1 & -1 \\ 2 & 3 & -3 \end{vmatrix} = -10 + 30 = 20 \neq 0.$$

Hence the vectors are linearly independent.

2. Let V be the space of all twice differentiable functions on [0, 1]. Find all $x \in V$, such that $x(t)$ and $x'(t)$ are linearly dependent.

Solution: $x'(t) = ax\,(b)$, $a \in F$, $a \neq 0$

$$\frac{dx}{dt} - ax = 0. \text{ Solving this equation, } x(t) = \lambda e^{at}.$$

3. Show that any subset of a linearly independent set is linearly independent.

Solution: Let S be a linearly independent set and $S' \subset S$.

Consider a linear relation $\sum_i a_i v_i = 0, v_i \in S', a_i \in F$.

This is also a linear relation of elements of S. So $a_i = 0$, for each i.

4. Under what conditions will $v_1 = a + ib$ and $v_2 = c + id$ be a basis of $V = C$ over reals?

Solution: $a + ib$ and $c + id$ must be linearly independent. The condition for this is

$$\begin{vmatrix} a & b \\ c & d \end{vmatrix} = ad - bc \neq 0.$$

5. Find a basis for the subspace of $\mathbb{R}^3$ defined by

$$V = \{x = (x_1, x_2, x_3) \mid x_1 + x_2 + x_3 = 0\}.$$

Solution: This is a plane in $\mathbb{R}^3$ and has a basis of 2 elements. $(1, -1, 0)$ and $(0, 1, -1)$ lie in V and are linearly independent. Hence it is a basis of V.

6. Let $V = \mathbb{R}^3$ and let $v_1 = (1, 1, 2)$, $v_2 = (0, 1, 3)$, $v_3 = (2, 4, 5)$ and $v_4 = (-1, 0, -1)$ be elements of V. Find a basis for the intersection of the subspace spanned by $\{v_1, v_2\}$ and $\{v_3, v_4\}$.

Solution: $V_1 = \{v_1, v_2\}$ is the plane $\begin{vmatrix} x & y & z \\ 1 & 1 & 2 \\ 0 & 1 & 3 \end{vmatrix} = 0$

i.e. $x - 3y + z = 0$

$V_2 = \{v_3, v_4\}$ is the plane $\begin{vmatrix} x & y & z \\ 2 & 4 & 5 \\ -1 & 0 & -1 \end{vmatrix} = 0$

i.e. $-4x - 3y + 4z = 0$

or $4x + 3y - 4z = 0$

$V_1 \cap V_2$ is the line defined by $x - 3y + z = 0$ and $4x + 3y - 4z = 0$. Solving these equations

$$\frac{x}{9} = \frac{y}{8} = \frac{z}{15}$$

So the vector $v = (9, 8, 15)$ is a basis of $V_1 \cap V_2$.

7. Let $V = \mathbb{R}$ be the space of real numbers considered as a vector space over Q. Does V have a basis of finite number of elements over Q? Why?

Solution: No. Since Q is countable and R is uncountable.

8. Let $\{v_1, v_2, ..., v_n\}$ be a maximal linearly independent set in V. Show that $\{v_1, v_2, ..., v_n\}$ is a basis of V.

Solution: It is sufficient to show that $\{v_1, ..., v_n\}$ generate V. Let $v \in V$ and consider $\{v, v_1, ..., v_n\}$. This set is linearly dependent as the given set is maximal linearly independent. So $av + \sum_i a_i v_i = 0$, not all a, a_i are zero. Now $a \neq 0$ or else $\{v_1 ... v_n\}$ will be linearly dependent.

So $v = -1/a\{a_1v_1 + ... + a_nv_n\}$ *i.e.* $\{v_1 ... v_n\}$ generate V.

9. Let V be a vector space with basis $\{v_1, v_2, ..., v_n\}$. Show that the following sets also form a basis of V.

(*i*) $v_1, v_2, ..., v_{i-1}, v_i + av_j, v_{i+1}, ... v_j, ..., v_n, i \neq j$

(*ii*) $av_1, ..., av_n, a \in F, \quad a \neq 0.$

Solution: (*i*) The given set is a generating set. Let $v \in V$

and write
$$v = \sum_{i=1}^{n} b_i v_i$$
$$= \sum_{k=1}^{i-1} b_k v_k + b_i (v_i + av_j) + b_{i+1} v_{i+1}$$
$$+ ... (b_j - ab_i) v_j + ... + b_n v_n.$$

It is also linearly independent because, consider a linear relation
$$\sum_{k=1}^{i-1} \lambda_k v_k + \lambda_i (v_i + av_j) + \sum_{k=i+1}^{n} \lambda_k v_k = 0.$$

Collecting coefficients and equating coefficients of all v_k to zero we have
$$\lambda_k = 0, k \neq j$$
$$a\lambda_i + \lambda_j = 0.$$

This implies λ_j is also zero.

Thus the set is a basis of V.

(*ii*) The given set is a generating set for V.

Let $v \in V$ and write
$$v = \sum_{i=1}^{n} b_i v_i$$
$$= \sum_{i=1}^{n} \left(\frac{b_i}{a}\right) av_i.$$

It is also linearly independent, for consider a linear relation
$$\sum_{k=1}^{n} \lambda_k av_k = 0.$$

They $\lambda_k a = 0$ for all k and since $a \neq 0$, $\lambda_k = 0$ for all k. Thus the set is a basis of V.

10. Let $F = Z_5$. Give two different sets of basis of $V = F^3$ over F.

Solution: (1, 0, 0), (0, 1, 0) and (0, 0, 1) is a basis. (1, 1, – 1) (2, 2, 0) and (– 1, 0, 1) is also a basis.

11. Let W be a subspace of V. Show that there exists another subspace W' of V such that $V = W \oplus W'$. Show further W' is not unique.

Solution: Choose a basis $\{v_1, ..., v_m\}$ of W and complete it to a basis $\{v_1, v_2, ..., v_m, v_{m+1}, ..., v_n\}$ of V. Let $W' = \{v_{m+1}, ..., v_n\}$. Then, clearly, $V = W \oplus W'$. W' is not unique as the following example shows:

$V = \mathbb{R}^2$
$$W = \{(x, 0) \mid x \in \mathbb{R}\}$$
$$W' = \{(0, y) \mid y \in \mathbb{R}\}$$

or $\qquad W' = \{(x, y) \mid y = \lambda x, \lambda \in \mathbb{R}, \lambda \neq 0\}.$

In both cases $\qquad V = W \oplus W'.$

12. Let V_1, V_2 and V_3 be subspaces of V. Show that

$$\dim (V_1 + V_2 + V_3) \leq \dim V_1 + \dim V_2 + \dim V_3 - \dim (V_1 \cap V_2) - \dim (V_2 \cap V_3) - \dim (V_1 \cap V_3) + \dim (V_1 \cap V_2 \cap V_3).$$

Solution: $\dim (V_1 + V_2 + V_3)$

$= \dim V_1 + \dim (V_2 + V_3) - \dim (V_1 \cap (V_2 + V_3))$

$= \dim V_1 + \dim V_2 + \dim V_3 - \dim (V_2 \cap V_3) - \dim (V_1 \cap (V_2 + V_3))$ (*)

Now $\qquad V_1 \cap (V_2 + V_3) \supset V_1 \cap V_2 + V_1 \cap V_3$

So $\dim (V_1 \cap (V_2 + V_3)) \geq \dim (V_1 \cap V_2) + \dim (V_1 \cap V_3) - \dim (V_1 \cap V_2 \cap V_3)$.

Substituting this in the above equality (*) we have the required inequality.

13. Let $K \subset F \subset E$ be fields such that F as a vector space over K has dimension n, E as a vector space over F has dimension m. Show that E as a vector space over K has dimension mn.

Solution: Let $\{v_1 \ldots v_n\}$ be a basis of F over K and $\{w_1 \ldots w_m\}$ a basis of E over F. Then the set $\{v_i w_j\}$ $1 \leq i \leq n$, $1 \leq j \leq m$ is a basis of E over K. Clearly it is a generating set.

If $w \in E$, $\qquad w = \sum_1^m a_i w_i,\ a_i \in F$.

Write each $\qquad a_i = \sum_{j=1}^m b_{ij} v_j,\ b_{ij} \in K$.

Substituting in w we have $\qquad w = \sum_{i,j} b_{ij} w_i v_j,\ b_{ij} \in K$.

Thus the set generates E over K. The set is also linearly independent. Consider a relation $\sum_{i,j} b_{ij} w_i v_j = 0$, $b_{ij} \in K$. Write this as $\sum_i (\sum_j b_{ij} v_j)\ w_i = 0$. Since $\{w_1 \ldots w_m\}$ are linearly independent we have $\sum_j b_{ij} v_j = 0$ for each i. Since $\{v_1 \ldots v_n\}$ are linearly independent $b_{ij} = 0$ for each j. Thus the set $\{v_i w_j\}$ is linearly independent and hence a basis of E over K.

3.4 LINEAR TRANSFORMATIONS

1. Let $V = R^2$. Show that rotation through an angle θ is a linear transformation of V in V. Is this linear transformation non-singular? Why?

Solution: The rotation map $T : \mathbb{R}^2 \to \mathbb{R}^2$ is given by

$$x' = x \cos \theta + y \sin \theta$$
$$y' = -x \sin \theta + y \cos \theta$$

Taking values on the standard basis (1, 0) and (1, 0) we find the matrix of T is

$\begin{pmatrix} \cos\theta & \sin\theta \\ -\sin\theta & \cos\theta \end{pmatrix}$. T is linear and non-singular as this matrix has an inverse

$$\begin{pmatrix} \cos\theta & -\sin\theta \\ \sin\theta & \cos\theta \end{pmatrix}.$$

2. Let V be the vector space of all twice differentiable functions in [0, 1]. Show that the mappings $T_1 : V \to V$ and $T_2 : V \to V$ defined by $T_1(f) = f'$ and $T_2(f) = xf$ are linear. Show further that $T_1T_2 - T_2T_1 = Id.$

Solution: T_1 is linear because $(f + g)' = f' + g'$ and $(af)' = af'$. T_2 is linear because $x(f + g) = xf + xg$ and $x(af) = axf$

$$(T_1T_2 - T_2T_1)(f) = T_1(xf) - T_2(f')$$
$$= f + xf' - xf' = f.$$

So $\quad T_1T_2 - T_2T_1 = Id.$

3. Let V and W be spaces over F such that $\dim V \geq \dim W$. Show that there exists a linear transformation of V onto W. Is the converse true? Why?

Solution: Choose a basis $\{v_1 \ldots v_n\}$ of V and $\{w_1 \ldots w_m\}$ of W with $n \geq m$. Define a map $T : V \to W$ by $T(v_i) = w_i$, $1 \leq i \leq m$ and $T(v_j) = 0, j > m$. T can be extended linearly and is a linear map onto W. Conversely let $T : V \to W$ be a map of V onto W.

Then $\quad V/\text{Ker } T \approx W$ and $\dim W = \dim V - \dim(\text{Ker } T) \leq \dim V.$

4. Show that a linear transformation $T : V \to V'$ is (1, 1) if and only if it maps a linearly independent set in V to a linearly independent set in V'.

Solution: Let T be (1, 1) and let $v_1, \ldots v_n$ be linearly independent in V. Let $T(v_i) = w_i$ $1 \leq i \leq n$. If $w_1 \ldots w_n$ are linearly dependent, then $\sum_j a_i w_i = 0$, not all a_i are zero.

Then $v = \Sigma a_i v_i \neq 0$, and $T(v) = 0$, a contradiction.

Conversely, suppose that T maps any linearly independent set to a linearly independent set. Let $v = \sum_j a_i v_i$ and $T(v) = 0$. Then $\Sigma a_i w_i = 0$. Since $\{w_1 \ldots w_n\}$ are linearly independent, $a_i = 0$, for each i. Then $v = 0$. Hence T is (1, 1).

5. Let $T : V \to V'$ be a linear transformation. If $\{v_1, v_2, \ldots, v_n\}$ are elements of V such that $T(v_1), T(v_2), \ldots, T(v_n)$ are linearly independent, show that $\{v_1, v_2, \ldots, v_n\}$ are linearly independent.

Solution: Suppose $\{v_1 \ldots v_n\}$ are linearly dependent. Then $\sum_i a_i v_i = 0$, not all a_i are zero. Then $\sum_i a_i T(v_i) = 0$. This implies $a_i = 0$ for all i, a contradiction.

6. Let $T : V \to V'$ be a linear transformation which is not "onto". Show that there exists some $v \in V$, $v \neq 0$, such that $T(v) = 0$.

Solution: $V/\text{Ker } T \approx T(V) \not\subseteq V$ as T is not onto.

$$\dim V - \dim (\text{Ker } T) < \dim V$$

Hence $\dim (\text{Ker } T) > 0$

So there exists $v \neq 0$, $v \in \text{Ker } T$, *i.e.* $T(v) = 0$.

7. Show that $V = F^{(n)}$ is isomorphic to $W = F^{(m)}$ if, and only if, $m = n$.

Solution: If $m = n$, clearly $F^{(m)} = F^{(n)}$.

Conversely if $F^{(m)} \approx F^{(n)}$, $\dim F^{(m)} = \dim F^{(n)}$ *i.e.* $m = n$.

8. Let V_1 and V_2 be subspaces of V. Show that

$$\frac{V_1 + V_2}{V_2} \approx \frac{V_1}{V_1 \cap V_2}.$$

Solution: Consider the map $T : V_1 \to \dfrac{V_1 + V_2}{V_2}$, given by $T(v) = v + V_2$, $v \in V_1$.

T is linear and "onto" with $\text{Ker } T = V_1 \cap V_2$.

Hence $\dfrac{V_1}{V_1 \cap V_2} \approx \dfrac{V_1 + V_2}{V_2}$.

9. Let V_1 and V_2 be subspaces of V such that $V_2 \subset V_1$. Show that $\dfrac{V/V_2}{V_1/V_2} \approx \dfrac{V}{V_1}$.

Solution: Consider the map $T : \dfrac{V}{V_2} \to \dfrac{V}{V_1}$ given by $T(v + V_2) = v + V_1$. T is well-defined because $V_2 \subset V_1$ and is linear, "onto". $\text{Ker } T = V_1/V_2$.

Hence by isomorphism theorem, $\dfrac{V/V_2}{V_1/V_2} \approx \dfrac{V}{V_1}$.

10. Give an example to show that the set of all non-zero linear transformations of V in V is not a group for composition.

Solution: A linear map $T : V \to V$ need not be invertible unless $\text{Ker } T = (0)$. Let $V = R^2$ and $T : V \to V$ defined by

$$T(x, y) = (x + y, x + y).$$

T has no inverse.

11. Show that the set $L(V, V)$ of all linear transformations of V in V is a ring, for the usual addition and composition of linear transformations. What are the units of this ring?

Solution: Clearly $L(V, V)$ is an abelian group with zero linear map as zero element and inverse of T is $(-T)$. Also composition is associative and distributes over addition, *i.e.*

$$T_1 (T_2 + T_3) = T_1 T_2 + T_1 T_3$$

There $L(V, V)$ is a ring. The units in this ring are the invertible linear maps of V in V.

12. Let $V = V_1 \oplus V_2$ be a direct sum of subspaces V_1 and V_2. Show that the maps $p_1 : V \to V_1$ and $p_2 : V \to V_2$ defined by $p_1(v) = v_1$ and $p_2(v) = v_2$, where $v = v_1 + v_2$, $v_1 \in V_1$, $v_2 \in V_2$, are linear maps of V onto V_1 and V_2 respectively. What are the kernels of p_1 and p_2 ? Show that the $V/V_1 \approx V_2$ and $V/V_2 \approx V_1$.

Solution: If $\quad v = v_1 + v_2 \quad$ and $\quad v' = v'_1 + v'_2$

then $\quad v + v' = v_1 + v'_1 + v_2 + v'_2$

so that $p_1(v + v') = v_1 + v'_1 = p_1(v) + p_1(v')$.

Similarly $\quad p_1(av) = ap_1(v)$. Thus p_1 is linear.

Similarly p_2 is linear. p_1 is "onto" because v_1 is the image of $(v_1, 0)$. Similarly p_2 is also "onto". Ker $p_1 = V_2$ and Ker $p_2 = V_1$.

So, $\quad V/V_2 \approx V_1 \quad$ and $\quad V/V_1 \approx V_2$.

13. Let v_1 and v_2 be two elements of V such that $v_2 \neq av_1$ for any $a \in F, a \neq 0$. Show that the there exists some $f \in V^*$ such that $f(V_1) = 1$ and $f(V_2) = 0$.

Solution: Since $v_2 \neq av_1$ for any a, $\{v_1, v_2\}$ is a linearly independent set and can be completed to a basis of V. Define a linear map $f^* : V \to F$ by prescribing the value 1 on the basis element v_1 and 0 on all the other basis elements.

14. Let $f \in V^*$ and $g \in V^*$ be such that $f(v) = 0$ implies that $g(v) = 0$, $v \in V$. Show that $g = af$, for some $a \in F$.

Solution: If $g = 0$ choose $a = 0$. If $g \neq 0$, there exists some w such that $g(w) \neq 0$, by assumption, $f(w)$ is also $\neq 0$. Thus $g(w) = a f(w)$ for some $a \in F$. We claim that for any $v \in V$, $g(v) = a f(v)$. Let $w' \in V$ be such that $g(w') = a' f(w')$.

$$f(f(w')\, w - w' f(w)) = f(w') f(w) - f(w') f(w) = 0$$

So $$g(f(w')\, w - w' f(w)) = 0$$

i.e. $$f(w')\, g(w) - g(w') f(w) = 0$$

$$f(w')\, a f(w) - a' f(w') f(w) = 0$$

Since $$f(w) \neq 0 \quad \text{and} \quad f(w') \neq 0,\ a = a'.$$

Thus a is the same for all elements w,

i.e. $\quad g(v) = a f(v)$ for all $v \in V$, *i.e.* $g = a f$.

3.5 MODULES

1. Let N be a finitely generated submodule of M such that M/N is finitely generated. Show that M is finitely generated.

Solution: Let $N = \{y_1 \ldots y_n\}$ and $M/N = \{\bar{x}_1, \ldots, \bar{x}_m\}$ be generating sets. We claim that M is generated by $\{x_1, \ldots, x_m, y_1, \ldots, y_n\}$. Let $x \in M$ and consider $\bar{x} \in M/N$.

Write $$\bar{x} = \sum_{i=1}^{m} a_i \bar{x}_i,\ a_i \in R.$$

Then $x - \sum_i a_i x_i \in N$ and so

$$x - \sum_i a_i x_i = \sum_{j=i}^{n} b_j y_j .$$

Hence $$x = \sum_i a_i x_i + \sum_j b_j y_j.$$

Thus $\{x_1, \ldots, x_m, y_1, \ldots, y_n\}$ generates M.

2. Show by an example that an increasing union of finitely generated submodules of M need not be finitely generated.

Solution: Let C_p^n be the group of p^nth roots of unity. It is a finite Z-module. Let $G = \bigcup_{n=0}^{\infty} C_p^n$. Then we claim that G is not a finitely generated Z-module.

Suppose $G = \{x_1 \ldots x_m\}$ is a finite generating set. Then each x_i belongs to some C_p^n and we can find N such that all $x_i \in C_p^N$ *i.e.* $G \subset C_p^N$, and hence $G = C_p^N$. This is a contradiction.

3. A submodule N of M is called a direct summand of M if $M = N \oplus N'$ for some submodule N' of M. If $K \subset N \subset M$ are submodules, show the following:

(*i*) If N is a direct summand of M, N/K is a direct summand of M/K.

(*ii*) If K is a direct summand of N, and N is a direct summand of M, then K is a direct summand of M.

(*iii*) If K is a direct summand of M, then K is a direct summand of N. If further N/K is a direct summand of M/K, then N is a direct summand of M.

Solution:

(*i*) Let $M = N \oplus N'$ and $K \subset N$.

Now $M/K = N/K + N'/K$ as $M = N + N'$.

Also $N/K \cap N'/K = N \cap N'/K = \{0\}$ as $N \cap N' = \{0\}$

So $M/K = N/K \oplus N'/K$.

(*ii*) Let $N = K \oplus K'$.

Then $M = N \oplus N' = K \oplus K' \oplus N'$.

So K is a direct summand of M.

(*iii*) Let $M = K \oplus K'$. Since $N \supset K$,

$$N = K \oplus K' \cap N.$$

To show this, clearly

$$N = K + K' \cap N.$$

Also $K \cap K' \cap N \subset K \cap K' = \{0\}$.

So $N = K \oplus K' \cap N$.

Now $M/K \approx K'$ and $N/K \approx K' \cap N$. Since N/K is a direct summand of M/K, $K' \cap N$ is a direct summand of K'. Write $K' = K' \cap N \oplus L$.

Then $M = K \oplus K' = K \oplus K' \cap N \oplus L = N \oplus L$.

Thus N is a direct summand of M.

4. Let N and K be submodules of M with $I = \text{Ann}(N)$ and $J = \text{Ann}(K)$. Show that $\text{Ann}(N \cap K) \supset I + J$ and give an example to show that the inclusion is strict.

Solution: Let $a \in I$ and $b \in J$ so that $aN = 0$ and $bK = 0$. So $a(N \cap K) = 0$ and $b(N \cap K) = 0$. So that $a + b \in \text{Ann}(N \cap K)$ *i.e.* $I + J \subset \text{Ann}(N \cap K)$. To show that the inclusion is strict, let $R = Z$, and $M = Z \times Z$, $N = Z \times \{0\}$, $K = \{0\} \times Z$.

Then $\qquad I = 0 = J \quad \text{and} \quad N \cap K = \{0\} \times \{0\}$

so that $\text{Ann}(N \cap K) = Z \times Z$

Thus $\qquad I + J \not\supset \text{Ann}(N \cap K)$.

5. Let N and K be submodules of M. Define the quotient $(N : K) = \{a \in R \mid aK \subset N\}$. Show that $(N : K)$ is a left ideal of R and $(N : K) = \text{Ann}\left(\dfrac{N+K}{N}\right)$.

Solution: If $a, b \in (N : K)$ then $aK \subset N$ and $bK \subset N$ so that $(a + b)K \subset N$. Also for any $x \in R$, $xaK \subset N$.

Hence it is a left ideal.

$$\text{Now} \qquad \text{Ann}\left(\frac{N+K}{N}\right) = \{a \in R \mid a(N+K) \subset N\}$$

$$= \{a \in R \mid aK \subset N\}$$

$$= (N : K)$$

6. Show that any finitely generated left R-module can be expressed as the quotient of a free R-module F of finite rank.

Solution: Let $\{x_1, \ldots, x_n\}$ be a generating set for M.

Let F be a free left module of rank n with basis $\{e_1, e_2, \ldots e_n\}$. Consider the map $\theta : F \to M$ defined by $\theta(e_i) = x_i$, $1 \le i \le n$ and extended R-linearly,

i.e. $\qquad \theta\left(\sum_i a_i e_i\right) = \sum a_i x_i.$

Then θ is a well defined left R-module homomorphism which is 'onto' since M is generated by $\{x_1 \ldots x_n\}$. Hence $M \approx F/Ker\,\theta$.

7. A module M is called simple if its only submodules are 0 and M. Show that a simple R-module is cyclic.

Solution: Let $M \neq 0$ and choose $a \in M$, $a \neq 0$. Then Ra is a submodule of M containing $a \neq 0$ and hence $Ra = M$. Hence M is cyclic.

8. Let M be a simple R-module. Show that any non-zero homomorphism $f : M \to M$ is an isomorphism.

Solution: Let $K = \text{Ker}(f)$. Since M is simple $K = 0$ or $K = M$. Since $f \neq 0$, K cannot be M. So $K = 0$, *i.e.* f is (1, 1). Also $f(M) \subset M$ is a non-zero submodule of M and $f(M) = M$. Thus f is "onto" and is an isomorphism.

9. Show that the abelian group Q as a Z-module is not finitely generated.

Solution: Suppose Q is generated by $\{x_1, \ldots, x_n\}$ and let $x_i = p_i/q_i$. Then any Z-linear combination of $x_1, \ldots, x_n$ has denominator a divisor of $q_1q_2, \ldots, q_n$ and so the rational number $\dfrac{1}{q_1 q_2 \cdots q_n + 1}$ is not in this combination. This is a contradiction to the assumption that $\{x_1, \ldots, x_n\}$ generates Q.

10. Show that if $M = M_1 \oplus M_2$ then $M/M_1 \approx M_2$ and $M/M_2 \approx M_1$.

Solution: Consider the projection map $p_1 : M \to M_1$ defined by $p_1(x) = x_1$, where $x = x_1 + x_2, x_1 \in M_1, x_2 \in M_2$. Then p_1 is an R-homomorphism of M onto M_1 with kernel M_2. Hence $M/M_2 \approx M_1$.

Similarly the second projection $p_2 : M \to M_2, p_2(x) = x_2$ has kernal M_1 and induces an isomorphism $M/M_1 \approx M_2$.

11. Show that $M = Z/p^n$, p prime, $n \geq 1$ considered as a Z-module, cannot be expressed as a direct sum of two non-zero submodules. (Such a module is called *indecomposable.*)

Solution: Let $M = Z/Zp^n = M' \oplus M''$. Since M is cyclic both M' and M'' are cyclic abelian groups so that

$$M' = \frac{(\lambda)}{p^n} \quad \text{and} \quad M'' = \frac{(\mu)}{p^n}.$$

Since $\lambda \mid p^n, \lambda = p^i, 1 \leq i \leq n-1$, as $M' \neq 0$

and similarly $\mu = p^i, 1 \leq j \leq n-1$, as $M'' \neq 0$.

Let $k = \text{Max}\,(i, j) > 0$ since the decomposition is proper. Then the $x = p^k$ (mod p^n) lies in both M' and M'' and is non-zero. This is a contradiction.

12. Is Q as a Z-module, indecomposable? Why?

Solution: Suppose $Q = Q_1 \oplus Q_2$ is a proper decomposition.

Choose $q_1 \in Q_1$, and $q_2 \in Q_2$ both non-zero.

Let $q_1 = r_1/s_1$ and $q_1 = r_2/s_2$.

Then $s_1r_2q_1 \in Q_1$ and $s_2r_1q_2 \in Q_2$.

But $s_1r_2q_1 = s_2r_1q_2 = r_1r_2 \in Q_1 \cap Q_2$ a contradiction because $Q_1 \cap Q_2 = (0)$. Hence Q is indecomposable.

3.6 STRUCTURE THEOREM

1. Find all abelian groups of order 360.

Solution: $360 = 2^3 \times 3^2 \times 5$. There are 6 abelian groups of order 360 isomorphic respectively to

$Z_2^3 \times Z_3^2 \times Z_5, Z_2^3 \times Z_3 \times Z_3 \times Z_5, Z_2^2 \times Z_2 \times Z_3^2 \times Z_5,$

$Z_2^2 \times Z_2 \times Z_3 \times Z_3 \times Z_5,$

$Z_2 \times Z_2 \times Z_2 \times Z_3^2 \times Z_5,$

$Z_2 \times Z_2 \times Z_2 \times Z_3 \times Z_3 \times Z_5.$

2. For which positive integers n, is it true that the only abelian group of order n is cyclic? Why?

Solution: n should be a product of distinct primes for if n has a square prime factor p^2, G has a subgroup of order p^2 and one possibility for this subgroup is $Z_p \times Z_p$ which is not cyclic. This is not possible as G is assumed to be cyclic.

3. Let p and q be distinct primes. How does the number of abelian groups of order p^r compare with the number of abelian groups of order q^r $(r \geq 1)$?

Solution: They are the same, both being the number of partions of r.

4. Show that if R is a commutative ring such that every sub-module of a f-g. free R-module is free, then R must be a PID.

Solution: Since R is R-free, every ideal I is free, and rank $I \leq 1$. In particular I is principal $I = (a)$. Moreover a is torsion free *i.e.* Ann $(a) = (0)$. I being free, there does not exist $b \in R$, $b \neq 0$ with $ba = 0$. Since a is arbitrary this implies that R is a domain. Hence R is a PID.

5. Show that Q/Z is a torsion group which has only one subgroup of order n, for each $n \geq 1$, this subgroup being cyclic.

Solution: For each element $a + Z$ of Q/Z, write $a = p/q$.

Then $$q(a + Z) = p + Z = \overline{0}.$$

Thus every element of Q/Z is a torsion element. Given n, $(1/n)$ is a cyclic subgroup of order n. Let H be any subgroup of order n.

Let $\quad h \in H$, $h = p/q$, $(p, q) = 1$. Then $nh = \overline{0}$

i.e. $\quad np/q \in Z$ *i.e.* q/np. Since $(q, p) = 1$, $q \mid n$ *i.e.* $n = qq'$.

Then $$h = \frac{q'p}{q'q} = \frac{q'p}{n} \in \left(\frac{1}{n}\right)$$

Thus $H \subset (1/n)$ and since both groups are of order n, $H = (1/n)$.

6. Find a basis for the submodule of Z^3 generated by $(1, 0, -1)$, $(1, 2, 3)$, $(-1, 2, 1)$, $(1, 1, 0)$.

Solution: The basis is given by the three elements: $(1, 0, -1)$, $(1, 1, 0)$ and $(1, 2, 3)$ because

$$(-1, 2, 1) = -4\,(1, 0, -1) + 4\,(1, 1, 0) - (1, 2, 3).$$

Also $(1, 0, -1)$, $(1, 1, 0)$ and $(-1, 2, 1)$ are independent because

$$\lambda\,(1, 0, -1) + \mu\,(1, 1, 0) + \nu\,(-1, 2, 1) = 0 \text{ implies } \lambda = \mu = \nu = 0.$$

Moreover $(-1, 2, 1)$ cannot be expressed as a combination of any two of the above 3 elements.

7. Let $R = Z[x]$. Show that the ideal $I = (2, x)$ cannot be written as a direct sum of cyclic $Z[x]$ modules.

Solution: The ideal $I = (2, x)$ is not principal. For if it is generated by $f(x)$, 2 must be a multiple of $f(x)$, *i.e.* $f(x)$ must be constant polynomial. Then x cannot be a multiple of $f(x)$.

Suppose $$I = (f(x)) \oplus (g(x)) \oplus \ldots$$

The rank (I) must be at least 2, which is not possible as I, being a submodule of $Z[x]$, must have rank ≤ 1. Thus if I is expressible as a direct sum of cyclic modules there must be only one factor. We have already shown that I cannot be free on one generator.

8. Let M be a finitely generated module over a PID and N a submodule of M. Show that rank M = rank N + rank $\dfrac{M}{N}$.

Solution: Let $M = F \oplus R/(q_1) \oplus + \ldots + R/(q_t) = F \oplus T$ and $N = F' \oplus T'$, where F' is the free part and T' the torsion part.

Then $$\frac{M}{N} = \frac{F \oplus T}{F' \oplus T'} \approx \frac{F}{F'} \oplus \frac{T}{T'}$$

To prove this consider the homomorphism

$$\theta : M = F \oplus T \to F/F' \oplus T/T'$$

given by $$\theta(a + b) = \bar{a} + \bar{b}$$

θ is a homomorphism which is "onto" and Ker $\theta = F' \oplus T'$

Hence $$\frac{F \oplus T}{F' \oplus T'} \approx \frac{F}{F'} \oplus \frac{T}{T'}.$$

Hence $$\text{rank } \frac{M}{N} = \text{rank } \frac{F'}{F'} = \text{rank } F - \text{rank } F'$$
$$= \text{rank } M - \text{rank } N.$$

9. Let M be a torsion module over a PID. Show that M is simple if and only if M is cyclic with prime exponent.

Solution: Assume M is cyclic $M = (a)$ with prime exponent p. *i.e.* $M \approx R/(p)$. Let $M' \subset M$ be a submodule. Then $M' \approx (q)/(p)$ since R is a PID. In particular, $q \mid p$, and since p is a prime, $q = 1$ or $q = p$, *i.e.* $M' = M$ or $M' = \{0\}$. Hence M is simple.

Conversely assume that M is simple. In particular, M is cyclic as for any $a \in M$, $a \neq 0$, $Ra = M$. Let the exponent of a be $n \in R$ and suppose it is not prime. Then $n = kl$, $(k, l) = 1$, proper divisors and

$$M \approx \frac{R}{(n)} \approx \frac{R}{(k)} \oplus \frac{R}{(l)}$$

is a proper decomposition contradicting the fact that M is simple.

10. Let M be a torsion module over a PID. Show that M is indecomposable if, and only if, M is cyclic with prime power exponent.

Solution: Since M is a torsion module

$$M = R/(q_1) \oplus \ldots \oplus R/(q_t)$$

and since it is indecomposable, $t = 1$ so that $M \approx R/(q_1)$ *i.e.* M is cyclic. If q_1 has distinct prime factors, $R/(q_1)$ will be decomposable as in the previous problem. So q_1 has a single prime factor, *i.e.* it is the power of a prime. So M is cyclic with prime power exponent. Conversely assume that M is cyclic with prime power exponent. Then $M \approx R/(p^k)$. By structure theorem

$$M \approx \frac{R}{(p^{k_1})} \oplus \frac{R}{(p^{k_2})} \oplus \ldots$$

$$k_1 \leq k_2 \leq \ldots$$

Since M is cyclic, by the uniqueness of the decomposition, there is only one term in the above decomposition, so that M is indecomposable.

11. Show that a finite abelian group which is not cyclic contains a subgroup isomorphic is $Z_p \oplus Z_p$ for some prime p.

Solution: Since G is finite abelian, by structure theorem

$$G \approx Z_{q1} \oplus \ldots \oplus Z_{qk}$$

All these q_i's cannot be mutually co-prime for otherwise G will be cyclic. So there is a prime p which divides two distinct q_i's say q_i and q_j, $(i \neq j)$. Then $Z_{qi} \oplus Z_{qj}$ contains $Z_p \oplus Z_p$ and hence G contains $Z_p \oplus Z_p$.

12. Let G_1 and G_2 be finite abelian groups such that G_1 is isomorphic to a subgroup of G_2 and G_2 is isomorphic to a subgroup of G_1. Show that G_1 and G_2 are isomorphic.

Solution: Clearly both G_1 and G_2 have the same number of elements n, both being finite. We first prove result for the case $n = p^k$.

G_1 will be isomorphic to a direct product of cyclic groups of orders

$$p^{\alpha_1}, p^{\alpha_2}, \ldots, p^{\alpha_t}, \ \alpha_1 + \alpha_2 + \ldots \alpha_t = k$$

and G_2 will be isomorphic to a direct product of cyclic groups of orders

$$p^{\beta_1}, p^{\beta_2}, \ldots, p^{\beta_s}, \ \beta_1 + \ldots \beta_s = k$$

Since G_1 is isomorphic to a subgroup of G_2 and vice versa the set $\{\alpha_1 \ldots \alpha_t\}$ will be a subset of $\{\beta_1 \ldots \beta_s\}$ and vice versa. Hence the two sets are identical and G_1 will be isomorphic to G_2.

In the general case, let $n = p_1^{\alpha_1} \ldots p_k^{\alpha_k}$

$$G_1 \approx H_1 \times \ldots \times H_k \quad \text{where } H_i = \{x \in G_1 \mid p_i^{\alpha_i} x = 0\}$$

and $$G_2 \approx L_1 \times \ldots \times L_k \quad \text{where } L_i = \{x \in G_2 \mid p_i^{\alpha_i} x = 0\}$$

Since G_1 is isomorphic to a subgroup of G_2 and vice versa, H_i will be isomorphic to a subgroup of L_i and vice versa $(1 \leq i \leq k)$.

By the first part

$$H_i \approx L_i, \ 1 \leq i \leq k \text{ and so } G_1 \approx G_2.$$

13. Let G be a finite abelian group of order n and $d \mid n$. Show that G has a subgroup of order d.

Solution: $G = Z_{q1} \oplus \ldots \oplus Z_{q_k}$

with $n = q_1 \ldots q_k$. If $d \mid n$, then

$$d = q_1 \ldots q_k \text{ with } q_1 \mid q_1, \ldots q_k \mid q_k$$

Then $H = Z_{q_1'} \oplus \ldots \oplus Z_{q_k'}$ is the required subgroup.

14. Find all abelian groups of order 90,000.

Solution : $90{,}000 = 2^4 \times 3^2 \times 5^4$

So the number of abelian groups of order 90,000 are

$$p(4) \times p(2) \times p(4) = 5 \times 2 \times 5 = 50.$$

These can be enumerated the groups

$Z_2^4 \times Z_3^2 \times Z_5^4$, $Z_2^3 \times Z_2 \times Z_3^2 \times Z_5^4$, $Z_2^2 \times Z_2 \times Z_2 \times Z_3^2 \times Z_5^4$,

$Z_2^2 \times Z_2^2 \times Z_3^2 \times Z_5^4$, $Z_2 \times Z_2 \times Z_2 \times Z_2 \times Z_3^2 \times Z_5^4$, and so on.

CHAPTER 4

Fields

4.1 FIELD EXTENSIONS

1. Find the degrees of the following field extensions:

(*i*) $(Q(\sqrt[3]{2}, \sqrt{3})/Q$, (*ii*) $Q(\sqrt{2}, \sqrt{3}, \sqrt{5})/Q$.

Solution:

(*i*) $$[Q(\sqrt[3]{2}, \sqrt{3}):Q] = [Q(\sqrt[3]{2}, \sqrt{3}):Q(\sqrt{3})][Q\sqrt{3}:Q]$$
$$= 3 \cdot 2 = 6$$

(*ii*) $[Q(\sqrt{2}, \sqrt{3}, \sqrt{5}):Q] = 2 \cdot 2 \cdot 2 = 8$

2. Show that a finite extension of prime degree is a simple extension.

Solution: Let K/F be of prime degree $p > 1$. Let $\alpha \in K - F$.

Then $F \subset F(\alpha) \subset K$ and deg (α) divides p.

Hence deg $\alpha = p$, *i.e.*, $K = F(\alpha)$.

3. Let K/F be an extension and L the set of all elements of K, algebraic over F. Is L algebraically closed? Why?

Solution: No. Take $F = Q$ and $K = R$. (reals.) Then L is not algebraically closed as the element α with $\alpha^2 = -1 \in L$ does not belong to L, since it does not belong to R.

4. Show that a finite field cannot be algebraically closed.

Solution: Let $F = [a_1, ..., a_p]$ be a finite field and consider the polynomial

$$f(x) = 1 + \prod_{i=1}^{p} (x - a_i) \in F[x].$$

This polynomial has no root in F, as none of the elements of F satisfy this polynomial.

5. Show that an extension K/F is algebraic if, and only if, every ring R with $F \subset R \subset K$ is a field.

Solution: Let K/F be algebraic and let $F \subset R \subset K$. If $\alpha \in R$, $\alpha^{-1} \in K$ and since K is algebraic, α^{-1} and hence α satisfies a polynomial

$$a_0 + a_1 \alpha + \ldots + a_n \alpha^n = 0, a_i \in F.$$

Multiplying by α^{-1}, we have

$$a_0 \alpha^{-1} = -(a_1 + a_2 \alpha + \ldots + a_n \alpha^{n-1}) \in R$$

and hence $\alpha^{-1} \in R$ as $a_0^{-1} \in F \subset R$. Thus R is a field.

Conversely assume that every ring R, such that $F \subset R \subset K$, is a field and let $\alpha \in K$. Take $R = F[\alpha]$. Since R is a field, $\alpha^{-1} \in R$.

Hence $\alpha^{-1} = f(\alpha)$ for some $f(x) \in F[x]$.

In particular, α satisfies $xf(x) - 1 \in F[x]$.

Hence α is algebraic over F.

6. Let K/F be an extension, $\alpha, \beta \in K$ algebraic over F with degrees m and n respectively. Show that if $(m, n) = 1$, then

$$[F(\alpha, \beta) : F] = mn.$$

Solution: $[F(\alpha, \beta) : F] = [F(\alpha, \beta) : F(\alpha)]\,[F(\alpha) : F] \leq mn$

Conversely $n \mid [F(\alpha, \beta) : F]$ and $m \mid F(\alpha, \beta) : F]$

and since $(m, n) = 1, mn \mid [F(\alpha, b) : F]$

In particular, $mn \leq [F(\alpha, \beta) : F]$.

Hence $[F(\alpha, \beta) : F] = mn$.

7. Prove that there exists a field K such that

(*i*) K is infinite.

(*ii*) K is algebraic over a finite field.

(*iii*) K is not algebraically closed. Justify.

Solution: Consider the finite field $F = Z_p$, p odd prime. The map $a \to a^2$ from F to F is not surjective as a and $-a$ get mapped "onto" the same element. Hence there exists some $a \in F$ which is not a square. Consider the polynomial $x^2 - a \in F[x]$ and adjoin a root α to F to get $F(\alpha) = F_{p^2}$ a field with p^2 elements. Continue the process to get an infinite sequence of fields

$$F = F_p \not\subseteq F_{p^2} \not\subseteq F_{p^4} \subset \ldots$$

Then $K = \bigcup_{i=1}^{\infty} F_{p^{2^i}}$ is an infinite field and is an algebraic extension of F. Also K is not algebraically closed as the polynomial $x^3 - 1 \in F[x] \subset K[x]$ does not have all the roots in K, as the minimum polynomial of any element of K is of the type

$$x^{2^m} - a \in F[x].$$

8. An algebraic number α is called an algebraic integer if it satisfies a monic polynomial relation with coefficients in Z. Show that:

(*i*) For any algebraic number α, $n\alpha$ is an algebraic integer for some $n \geq 1$.

(*ii*) If a rational number is an algebraic integer, it must belong to Z.

Solution: (*i*) Since α is an algebraic number, α satisfies a relation

$$\alpha^n + a_1\alpha^{n-1} + \ldots + a_n = 0, \quad a_i \in Q.$$

Write $a_i = p_i/q_i, p_i, q_i \in Z$ and let $q = \pi q_i$.

Multiplying by q^n we have $\beta = \alpha q$ satisfies

$$\beta^n + b_1\beta^{n-1} + \ldots + b_n = 0, \quad b_i \in Z.$$

Hence β is an algebraic integer

(*ii*) Let $\alpha = p/q$ be an algebraic integer, $(p, q) = 1$. Then α satisfies

$$\alpha^n + a_1\alpha^{n-1} + \ldots + a_n = 0, \quad a_i \in Z$$

Then $$p^n/q^n + a_1 \frac{p^{n-1}}{q^{n-1}} + \ldots = 0$$

Multiplying by q^n, we have

$$p^n + b_1 q + \ldots + b_n q^n = 0, \quad b_i \in Z$$

We claim that q is a unit for if it has an irreducible factor, the same factor divides p^n by the above relation and hence divides p. This contradicts the assumption that $(p, q) = 1$. Hence $\alpha = p/q \in Z$.

9. Let $K = F(\alpha)$, where α is algebraic over F with odd degree. Show that $K(\alpha) = K(\alpha^2)$.

Solution: Consider the extensions $K \subset K(\alpha^2) \subset K(\alpha)$.

If $K(\alpha) \neq K(\alpha^2)$ then $[K(\alpha) : K(\alpha^2)] = 2$ and hence $[K(\alpha) : K]$ is even, a contradiction.

10. Using the fact that $x^n - p \in Z[x]$ is irreducible, p prime, $n > 1$, show that for every n, there is an algebraic element of degree n over Q. Deduce that $\bar{Q}/Q$ is not a finite extension.

Solution: $x^n - p \in Q[x]$ is irreducible if p is prime and $(n, p) = 1$. Choose such a prime and adjoin a root of this polynomial to Q. This element is algebraic of degree n over Q. Since all algebraic elements line in $\bar{Q}$ and n is arbitrary $[\bar{Q} : Q]\ \infty$.

11. Let K/F be an extension, and L_1 and L_2 subfields of K, which are finite extensions of F. Let $L = L_1L_2$ be the subfield of K generated by L_1 and L_2, *i.e.*, the smallest subfield of K containing $L_1 \cup L_2$. Show that L/F is finite and

$$[L : F] \leq [L_1 : F]\,[L_2 : F].$$

Solution: Let $[L_1 : F] = m$ and $[L_2 : F] = n$.

If $L_1 = F(\alpha_1, \ldots, \alpha_t)$, then $L = L_1L_2 = L_2(\alpha_1 \ldots \alpha_t)$

Now $[L : L_2] \leq [L_1 : F]$ because the degrees of $\alpha_1 \ldots \alpha_t$ over L_2 are less than or equal to their degree over F.

Since $$[L_1 : F] < \infty,$$

$$[L : F] = [L : L_2]\,[L_2 : F] \text{ is finite}$$

and $$[L : F] = [L : L_2]\,[L_2 : F] \leq [L_1 : F]\,[L_2 : F]$$

12. Let K/F be a finite extension. If for any two subfields L_1 and L_2 of K containing F, either $L_1 \subset L_2$ or $L_2 \subset L_1$, show that K is a simple extension of E.

Solution: Write $K = F(\alpha_1, \ldots, \alpha_t)$. For the fields $F(\alpha_1)$, $F(\alpha_2)$ …, $F(\alpha_t)$ we have either $F(\alpha_i) \subset F(\alpha_j)$ or $F(\alpha_j) \subset F(\alpha_i)$. Hence for some k, $F(\alpha_i) \subset F(\alpha_k)$ for all i, *i.e.*, $K \subset F(\alpha_k)$, *i.e.*, $K = F(\alpha_k)$. Hence K is a simple extension.

4.2 NORMAL EXTENSIONS

1. Find the degree of the splitting fields over Q of the polynomials

(*i*) $x^4 - 2$ (*ii*) $x^4 + 1$

Solution:(*i*) Splitting field of $x^4 - 2$ is $Q(\sqrt[4]{2}, i)$ and is of degree 8 over Q.

(*ii*) Splitting field of $x^4 + 1$ is a $Q(\sqrt{i})$ and is of degree 4 over Q.

2. Let α be a root of

$$f(x) = x^3 + x^3 + 1 \in Z_2[x].$$

Show that $Z(\alpha)$ is a splitting field of $f(x)$.

Solution: $f(x)$ is irreducible over Z_2. Clearly α^2 is also a root of $f(x)$ and the other root is $1 + \alpha + \alpha^2$. All these lie in $Z(\alpha)$ and hence it is the splitting field.

3. Show that the splitting field of $x^p - 1 \in Q[x]$, p prime, is of degree $p - 1$ over Q.

Solution: If ρ is a primitive pth root of unity, splitting field is $Q(\rho)$ and ρ satisfies the irreducible polynomial

$$1 + x + x^2 + \ldots + x^{p-1} \in Q[x].$$

Hence $$[Q(\rho) : Q] = p - 1.$$

4. Show that the splitting field of a polynomial of degree n over F has degree utmost $\lfloor n$. Give examples for each of the following:

(*i*) Splitting field has degree less than $\lfloor n$,

(*ii*) Splitting field has degree $\lfloor n$.

Solution: Adjoint a root of $f(x)$ to F to get $F(\alpha)$

and $$[F(\alpha) : F] \le n = \deg f.$$

$f(x)$ can be written as

$$f(x) = (x - \alpha) \quad g(x)\ g(x) \in F(\alpha)[x]$$

and by induction its splitting field has degree utmost $(\lfloor n-1)$ over $F(\alpha)$. But this field is the same as the splitting field of $f(x)$ over F and has degree utmost $(\lfloor n-1) \cdot [F(\alpha) : F] \le \lfloor n$ over F

$$x^4 - 2 \in Q[x] \text{ has splitting field of degree } 8 < \lfloor 4.$$

$$x^3 - 2 \in Q[x] \text{ has splitting field of degree } 6 = \lfloor 3$$

5. Illustrate by an example that a normal extension of a normal extension need not be a normal extension.

Solution: Consider $Q(\sqrt[4]{2})/Q$. The extension $Q(\sqrt[4]{2})$ is normal over $Q(\sqrt{2})$ and $Q(\sqrt{2})$ is normal over Q as both are of degree 2. But $Q(\sqrt[4]{2})$ is not normal over Q as i is not in this field.

6. Let $f(x) \in F[x]$ be an irreducible polynomial and K/F a finite normal extension. Let $g(x), h[x] \in K[x]$ be two irreducible factors of $f(x) \in K[x]$. Show that there exists an F-automorphism σ of K such that $\sigma(g(x)) = h(x)$.

Solution: Let E/F be a splitting field of $f(x) \in F[x]$. Since K/F is normal, K is the splitting field of some $\theta(x) \in F[x]$. Then the composite extension $L = KE$ over F is the splitting field of the product $f(x)\,\theta(x) \in F[x]$ and contains both K and E. Let α be a root of $g(x)$ in L and β a root of $h(x)$ in L. Since $h(x), g(x)$ are irreducible factors of $f(x) \in K[x]$, α and β are roots of $f(x)$. Hence there exists an automorphism $\sigma : E \to E$ with $\sigma(\alpha) = \beta$. Extend this to an automorphism $\bar{\sigma} : L \to L$ and since $K \subset L$ is normal, $\bar{\sigma}(K) = K$. Thus $\bar{\sigma} : K \to K$ is an automorphism with $\bar{\sigma}(\alpha) = \beta$. In particular, $\bar{\sigma}$ maps the minimum polynomial of α over K, *i.e.*, $g(x)$ "onto" the minimum polynomial of β over K, *i.e.*, $h(x)$. Hence $\bar{\sigma}$ is the required automorphism.

7. Let $F = Z_p$, prime. Show that there is an irreducible polynomial of degree 2 over F. Use this polynomial to construct a field with p^2- elements.

Solution: The map $\theta = F \to F$, $\theta(a) = a^2$ is not injective as a and $-a$ get mapped "onto" the same element. Hence θ is not "onto". Choose $a \in F$ which is not a square in F. Then $x^2 - a \in F[x]$ is irreducible of degree 2 over F. If α is a root of this polynomial in some extension field, $Z_p(\alpha)$ is a field with p^2-elements.

8. Find all the conjugates of the following elements over Q.

(*i*) $\sqrt{1+\sqrt{2}}$ (*ii*) $\sqrt{2} + i$

Solution: (*i*) Conjugates of $\sqrt{1+\sqrt{2}}$ are

$$\sqrt{1+\sqrt{2}},\ -\sqrt{1+\sqrt{2}},\ \sqrt{1-\sqrt{2}} \quad \text{and} \quad -\sqrt{1-\sqrt{2}}$$

(*ii*) Conjugates of $\sqrt{2} + i$ are

$$\sqrt{2}+i,\quad \sqrt{2}-i,\quad -\sqrt{2}+i \quad \text{and} \quad -\sqrt{2}-i.$$

9. Show that if n is an integer which is not a perfect square and $\alpha = a + b\sqrt{n}$, a, $b \in Q$ is a root of a polynomial $f(x) \in Q[x]$, then $a - b\sqrt{n}$ is also a root of $f(x)$.

Solution: $(\alpha - a)^2 = nb^2$ and so α is of degree 2 over Q, *i.e.*, $[Q(\alpha) : Q] = 2$. The other root of the minimum polynomial of α is $a - b\sqrt{n}$ as $\alpha - a = \pm\, b\sqrt{n}$.

Since the minimum polynomial of α divides $f(x)$, $a - b\sqrt{n}$ is also a root of $f(x)$.

10. Consider $f(x) = x^3 + ax + b \in Q[x]$. Give suitable values of a, b so that the splitting field of $f(x)$ has degree 3 over Q.

Solution: The discriminant of the cubic is $\Delta = 4a^3 - 27b^2$. If $\Delta \neq 0$, the cubic is irreducible. If Δ is a perfect square, the splitting field will be of degree 3 over Q, otherwise if will be of degree 6. For $a = 3$, $b = 1$, Δ is a perfect square and hence the splitting field will be of degree 3 over Q.

11. What is the splitting field of $x^{p^n} - 1 \in Z_p[x]$, $(n \geq 1)$.

Solution: The polynomial $x^{p^n} - 1 = (x-1)^{p^n}$ *as* $p \left| \binom{p}{i} \right.$ for $i > 0$ and ch $Z_p = p$. Hence the only root of this polynomial is 1 and the splitting field is Z_p.

12. Let K_1/F, K_2/F be finite normal extensions such that $K_1 \subset L$, and $K_2 \subset L$ for some field L. Show that K_1K_2 is a normal extension of F.

Solution: K_1 is the splitting field of a polynomial $f_1(x)$ and K_2 is the splitting field of a polynomial $f_2(x)$. Then K_1K_2 is the splitting field of $f(x) = f_1(x) f_2(x)$. Hence K_1K_2 is a normal extension.

4.3 SEPARABLE EXTENSIONS

1. Give an example of a polynomial $f(x) \in Q[x]$ which has no root in Q but all of whose roots are of multiplicity 2 in some extension. Explain why this does not contradict the fact that Q is a perfect field?

Solution: $f(x) = x^4 + 2x^2 + 1 \in Q(x)$ has no root in Q but roots i, and $-i$ are multiplicity 2 in C. There is no contradiction because the polynomial is reducible.

2. Find α such that $Q(\alpha) = Q(\sqrt{2}, \sqrt{3})$.

Solution: $\alpha = \sqrt{2} + \sqrt{3}$ is of degree 4 over Q.

Hence $\quad Q(\sqrt{2} + \sqrt{3}) = Q(\sqrt{2}), (\sqrt{3})$.

3. Let ch $F = p > 0$ and $f(x) \in F[x]$ an irreducible polynomial. Show that $f(x) = g(x^{p^e})$, $(e \geq 0)$ for some $g(x)$ which is irreducible and separable. Deduce that every root of $f(x)$ occurs with the same multiplicity.

Solution: If $f(x)$ is separable, take $e = 0$ and $g(x) = f(x)$. Otherwise $f(x)$ is a polynomial in x^p. Choose $e \geq 1$ such that $f(x)$ is a polynomial in x^{pe} but not in

$x^{p^{e+1}}$. Then $f(x) = g(x^{p^e}) = g(y)$ and $g(y)$ is separable because otherwise $g(y)$ will be a polynomial in y^p, *i.e.*, $f(x)$ is a polynomial in $x^{p^{e+1}}$, a contradiction. Let $\beta_1 \ldots \beta_m$ be the roots of $g(y)$ in some extension field.

Then $$f(x) = g(y) = \prod_{i=1}^{m} (y - \beta_i) = \prod_{i=1}^{m} (x^{p^e} - \beta_i)$$

If α_i is a root of $f(x)$, $a_i^{p^e} = \beta_i$,

i.e., $$x^{p^e} - \beta_i = x^{p^e} - \alpha_i^{p^e} = (x - \alpha_i)^{p^e},$$

i.e., α_i is of multiplicity p^e, which is the same for all roots.

4. Show that an algebraic extension of a perfect field is perfect.

Solution: Let F be a perfect field and K and algebraic extension of F. If L *is* any algebraic extension of K, L/F is algebraic and hence L/F is separable. In particular, L/K is separable. Hence K is perfect.

5. Let ch $F = p > 0$. Show that an element α algebraic over F is separable if and only if $F(\alpha) = F(\alpha^p)$.

Solution: Assume α is separable are F. There $F(\alpha)/F(\alpha^p)$ is also separable. Also α is purely inseparable over $F(\alpha^p)$ and so $F(\alpha)/F(\alpha^p)$ is purely inseparable.

Hence $F(\alpha) = F(\alpha^p)$

Conversely assume that $F(\alpha) = F(\alpha^p)$

The $\alpha = f(\alpha^p)$. For some $f(x) \in F[x]$, *i.e.*, α satisfies $x - g(x)$, where $g(x)$ is purely a polynomial in x^p. It α were inseparable, its minimum polynomial is purely a polynomial in x^p and must divide $x - g(x)$, which is a contradiction.

6. Let ch $F = p > 0$, Let $F^{p^{-n}}$ denote all the p^nth roots of elements of F (in some algebraic closure of F) and $F^{p^{-\infty}} = \bigcup_{n=1}^{\infty} F^{p^{-n}}$. Show that

(*i*) For each n, $F^{p^{-n}}$ is isomorphic to F.

(*ii*) $F^{p^{-\infty}}$ is a perfect field and it is the smallest perfect field containing F.

(*iii*) F is perfect if, and only if, $F = F^{p^{-\infty}}$.

Solution: (*i*) Consider the map $\theta : F^{p^{-1}} \to F$ given by $\theta(a) = a^p$. Then θ is an isomorphism. Thus $F^{p^{-1}} \approx F$. By iteration $F^{p^{-n}} \approx F$.

(*ii*) Let $\alpha \in F^{p^{-\infty}}$ so that $\alpha \in F^{p^{-n}}$ for some n. Then a pth root β of α lies in $F^{p^{-n+1}} \subset F^{p^{-\infty}}$. Hence every element of $F^{p^{-\infty}}$ has a pth root in it. Hence $F^{p^{-\infty}}$ is perfect. Any perfect field containing F contains $F^{p^{-n}}$ for each n and hence contains $F^{p^{-\infty}}$.

(*iii*) Follows from (*ii*).

7. Examine whether $x^4 + x + 1 \in Q[x]$ is a separable polynomial.

Solution: The polynomial is separable as

$$f(x) = x^4 + x + 1 \quad \text{and} \quad f'(x) = 4x^3 + 1$$

are relatively prime and have no common root. Hence $f(x)$ has no multiple roots in extension of Q.

8. Let ch $F = p > 0$. Show that $F(x, y)$ is not a simple extension of $F(x^p, y^p)$.

Solution: Clearly $[F(x, y) : F(x^p, y^p)] = p^2$.

If $\theta \in F(x^p, y^p)$ then deg θ over $F(x^p, y^p)$ will be p, and hence cannot generate $F(x, y)$.

9. Let ch $F = p > 0$ and $f(x) = x^p - x - a \in F[x]$.

Show that $f(x)$ has no multiple roots and is irreducible if and only if $a \neq c^p - c$ for any $c \in F$.

Solution: $f'(x) = -1$ and hence $f(x)$ and $f'(x)$ have no common root. Hence $f(x)$ has no multiple roots. If $a = c^p - c$ for some $c \in F$

$$f(x) = x^p - x - c^p + c = (x - c)^p - (x - c)$$
$$= (x - c)[(x - c)^{p-1} - 1] \text{ is reducible.}$$

Conversely if $a \neq c^p - c$, $f(x)$ is irreducible. $f(x)$ has no root in F. Let α be a root of $f(x)$ in some extension field of F. Then $\alpha + i$, $0 \leq i \leq p - 1$ are all the roots. Let $g(x)$ be the minimum polynomial of α. The deg $g(x) > 1$ and $g(x) \mid f(x)$. The roots of $g(x)$ are $\alpha + i$, $0 \leq i \leq p - 1$. Thus $g(x) = f(x)$ and $f(x)$ is irreducible.

10. Let ch $F = p > 0$ and K/F a finite extension. For any $\alpha \in K$, show that either $\alpha^{p^n} \in F$ for some n or there exists some m with $\alpha^{p^m} \notin F$ and separable over F.

Solution: Let $f(x)$ be the minimum polynomial of α. Then $f(x) = g(x^{p^e})$ and $g(x)$ is irreducible and separable. Since $0 = f(\alpha) = g(\alpha^{p^e})$, either $\alpha^{p^e} \in F$ or $\alpha^{p^e} \notin F$ and is separable over F as its minimum polynomial is $g(x)$.

11. Let K/F be an extension and $\alpha \in K$ algebraic over F. α is called purely inseparable if its minimum polynomial has only one root in its splitting field (without counting multiplicity). Show that

(*i*) α is purely inseparable if, and only if, $\alpha^{p^n} \in F$ for some n.

(*ii*) If α is separable and purely inseparable over F, then $\alpha \in F$.

(*iii*) α is purely inseparable are F if, and only if, its minimum polynomial over F is of the type $x^{p^e} - a \in F[x]$.

Solution: Since the minimum polynomial of α has only one root, it is of the type

$$f(x) = (x - \alpha)^{p^e} = x^{p^e} - a, \quad a = \alpha^{p^e}.$$

This proves (*iii*). Also, since $f(\alpha) = 0$, $\alpha^{p^e} = a \in F$ and conversely. This proves (*i*).

(*ii*) If α is further separable, $e = 0$ $f(x) = x - a$, $\alpha = a \in F$.

12. Let ch $F = p > 0$ and K/F a finite extension. Show that

(*i*) If K/F is inseparable, then p divides $[K : F]$

(*ii*) If K/F is purely inseparable, then $[K : F] = p^m$ for some $m \leq 0$.

Solution: (*i*) Suppose $p \nmid [K : F]$. Then for every $\alpha \in K$, deg (α) is relatively prime to p and hence the minimum polynomial of α is separable. Hence K/F is a separable extension.

(*ii*) Suppose K/F is purely inseparable.

Then $$K = F(\alpha_1, \ldots, \alpha_n)$$

where each α_i is purely inseparable.

Since $[F(\alpha_1 \ldots \alpha_i); F(\alpha_1 \ldots \alpha_{i-1})] = p^{e_i}$,

$$[K : F] = p^{e_1 + \ldots + e_n} = p^m.$$

13. Let K/F be an algebraic extension and L the set of elements of K separable over F. Show that L is a subfield of K and:

(*i*) L/F is separable

(*ii*) K/L is purely inseparable.

Solution: If α and β are separable over F, $F(\alpha, \beta)/F$ is a separable extension. Hence $\alpha \pm \beta$, $\alpha \cdot \beta$ and α/β ($\beta \neq 0$) are separable over F. Hence L is a subfield of K. Clearly

(*i*) L/F is separable.

(*ii*) Let $\alpha \in K$. Then a^{p^n} is separable over F for some $n \geq 0$. Then $a^{p^n} \in L$. Hence K/L is purely inseparable.

14. Let K/F be a simple extension. Show that the number of fields L with $F \subset L \subset K$, including F and K is utmost 2^{n-1}, where $n = [K : F]$.

Solution: Suppose $K = F(\alpha)$ for some α. Let $f(x)$ be the minimum polynomial of α so that deg $f(x) = n$. Let $F \not\subseteq L \not\subseteq K$ and $g_L(x)$ the minimum polynomial of α over L. There $g_L(x) \mid f(x)$ and we get a map $L \to g_L(x)$ which is injective. Hence the number of intermediary fields is utmost equal to the number of proper factors of $f(x)$ in $K[x]$ which is utmost 2^{n-1} as deg $f(x) = n$.

15. Let K/F be a finite extension. Show that K is a simple extension of F if, and only if, there are a finite number of intermediary fields between F and K.

Solution: If K is a simple extension, by the above problem, there are a finite number of intermediary fields between F and K. Conversely assume that there are only a finite number of intermediary fields between F and K.

Let $\alpha, \beta \in K$ and consider $\alpha + \lambda\beta$, $\lambda \in F$.

Then $$F(\alpha + \lambda_1\beta) = F(\alpha + \lambda_2\beta) = L$$
for some $\lambda_1, \lambda_2 \in F, \lambda_1 \neq \lambda_2$.

Since $\alpha + \lambda_1\beta$ and $\alpha + \lambda_2\beta$ are in the same field, so is $(\lambda_1 - \lambda_2)\ \beta$, and hence $\beta \in L$ and so $\alpha \in L$. So $F(\alpha, \beta) = L$.

Thus $F(\alpha, \beta)$ is generated by one element. Since $K = F(\alpha_1, \ldots, \alpha_n)$, proceeding inductively, we get that K is generated by one element.

4.4 FINITE FIELDS

1. Find the generators of the cyclic group of non-zero elements of the following finite fields.
   (*i*) Z_{19}, (*ii*) Z_{29}.

   Solution: (*i*) 2 mod (19) is a generator for Z^*_{19}.

   (*ii*) 2 mod (29) is also a generator for Z^*_{29}.

2. Show that if $m \mid n$, $x^{p^m} - x$ divides $x^{p^n} - x$.

   Solution: Divide $x^{p^n} - x$ by $x^{p^m} - x$ by long division method. Since $m \mid n$, there will be no remainder left.

3. Let F be a finite field with q elements and $f(x) \in F[x]$ an irreducible polynomial. Show that $f(x)$ divide $x^{q^n} - x$ if, and only if, deg (f) divides n.

   Solution: Consider $x^{q^n} - x \in F[x]$, and let K be a splitting field of this polynomial. Then $[K : F] = n$. If f divides $x^{q^n} - x$, and α is a root of $f(x)$, $F(\alpha) \subset K$ and so deg $(\alpha) \mid [K : F] = n$, *i.e.*, deg $(f) \mid n$.

   Conversely assume that deg $(f) \mid n$ and let α be a root of $f(x)$ in some extension field and $K = F(\alpha)$. Then $[K : F] = m \mid n$ and α satisfies $x^{q^n} - x$, as $m \mid n$. Since $f(x)$ is the minimum polynomial of α, $f(x) \mid x^{q^n} - x$.

4. Let K/F be a finite extension, where F is a finite field. Show that the number of elements of K is a power of the number of elements of F.

   Solution: Let ch $F = p > 0$ so that F has $q = p^n$ elements. If $[K : F] = m$, K has q^m elements.

5. Let F be a finite field with p^n-elements. Show that the map $a \to a^p$ is an automorphism of F (called the Frobenius automorphism). Show that it has order n.

   Solution: $\theta : F \to F, \theta(\alpha) = a^p$ is clearly a homomorphism which is (1, 1) and hence "onto". Then θ is an isomorphism of F onto F.
   $$\theta^2(a) = a^{p^2} \ldots \theta^i(a) = a^{p^i}.$$

So that $\theta^n(a) = a^{p^n} = a = I(a)$. Then $\theta^n = I$.

Also for $i < n$, $\theta^i(a) = a^{p^i} \neq a$, for a generator a of the cyclic group F^* of order $p^n - 1$. Hence θ has order n.

6. Let ch $F = p > 0$. Give an example to show that the map $a \to a^p$ need not be an automorphism when F is infinite field.

Solution: Let $K = F(x)$, the field of rational functions in x over F, ch $F = p > 0$. Then ch $K = p > 0$. The map $\theta : K \to K$, $\theta(f) = f^p$ is an injective homomorphism which is not "onto" as $x \in K$ has no pre-image under θ, *i.e.* $f(x)^p = x$ is impossible for any rational function f. Then θ is not an automorphism of K.

7. Show that every irreducible polynomial $f(x) \in Z_p[x]$ is a divisor of $x^{p^n} - x$ for some n.

Solution: If deg $f = n$, thereby Problem 3, $f(x)$ divides $x^{p^n} - x$.

8. Investigate whether a finite field with the following number of elements exists and if so construct such a field.

(*i*) 625 (*ii*) 70.

Solution: $q = 625 = 5^4$ is power of a prime. Hence there is a field with 625 elements.

Consider Z_5 and $f(x) \in Z_5[x]$ defined by

$$f(x) = x^4 + x^2 + x + 1.$$

Then $f(x)$ is irreducible. Adjoin a root α of this polynomial to Z_5 to get a field $K = Z_5(\alpha) = Z_5[\alpha]$. Then α will be a generator of K^*. Addition and multiplication in K can be defined subject to the conditions.

$$\alpha^4 + \alpha^2 + \alpha + 1 = 0 \quad \text{and} \quad \alpha^{624} = 1.$$

(*ii*) There is no field with 70 elements as 70 is not the power of a prime.

9. Show that $x^{p^n} - x$ is the product of all monic irreducible polynomials of deg d dividing n.

Solution: Let f be a monic irreducible polynomial of deg d dividing n so that $f \mid x^{p^d} - x \mid x^{p^n} - x$ as $d \mid n$. Two such f's are relatively prime so that the product of all such f's divide $x^{p^n} - x$. Conversely any irreducible factor of $x^{p^n} - x$ is a polynomial of deg $d \mid n$ by Problem 3. Hence the result.

4.5 GALOIS EXTENSION

1. Verify the fundamental theorem of Galois theory for the splitting field of $x^4 - 2 \in Q[x]$.

Solution: Splitting field of $f(x) = x^4 - 2$ is $K = Q(\alpha, i)$, where α is a real root of $x^4 - 2$. The other roots are $\pm\alpha, \pm i\alpha$. Let $\sigma : K \to K$ be the automorphism defined by

$$\sigma(\alpha) = i\alpha, \sigma(i) = i.$$

Then σ is of order 4. Let τ be the automorphism such that $\tau(\alpha) = \alpha$, $\tau(i) = -i$. Then the Galois group G which is of order 8 is generated by σ and τ with relation $\tau\sigma = \sigma^3\tau$. The proper subgroups of G are.

(*i*) $\{1, \tau\}, \{1, \sigma^2\tau\}, \{1, \sigma^2\}, \{1, \sigma\tau\}, \{1, \sigma^3\tau\}$ all of order 2.

(*ii*) $\{1, \sigma^2, \tau, \sigma^2\tau\}, \{1, \sigma, \sigma^2, \sigma^3\}, \{1, \sigma^2, \sigma\tau, \sigma^3\tau\}$ all of order 4.

The corresponding fixed fields can be written down and the Galois correspondence of bijection between subgroups of G and subfields of K, holds.

2. Find the order of the Galois groups of the splitting field of $x^3 - 10 \in Q(\sqrt{-3})$.

Solution: The discriminant of the cubic $x^3 - 10$ is $\Delta = -2700$ which is a perfect square in $Q(\sqrt{-3})$. Hence the Galois group of the polynomial is of order 3 and isomorphic to A_3, (Problem 7 of 4.7).

3. Let ch $F = p > 0$ and $f(x) = x^p - x - a \in F[x]$, such that a is not of the form $a = b^p - b$, $b \in F$. Find $G(K/F)$, where K is a splitting field of $f(x)$.

Solution: The polynomial $f(x)$ is irreducible since $a \neq b^p - b$, $b \in F$. Let α be a root of $f(x)$ in some extension field, so that all the other roots are $\alpha + i, 1 \leq i \leq p - 1$. Then $K = F(\alpha)$ is the splitting field of $f(x)$. Since $[K : F] = p$, $G(K/F)$ is a cyclic group of order p.

4. Show that a finite field is a Galois extension of any of its subfields.

Solution: Let K be a finite field and F a subfield of K. Let $[K : F] = n$, and F contain q elements. Then K is the splitting field of $x^{q^n} - x \in F[x]$ and hence K/F is a normal extension. It is also separable as F is a perfect field. Hence K/F is a Galois extension.

5. Let K_1/F and K_2/F be Galois extensions. Show that the composite K_1K_2/F is a Galois extension with Galois group isomorphic to a subgroup of $G(K_1/F) \times G(K_2/F)$.

Solution: K_1K_2 is separable over K_2 as K_1 is separable over F. Since K_2/F is separable, K_1K_2/F is separable. Since K_1/F and K_2/F are normal, these are the splitting fields of polynomials $f(x) \in F(x)$ and $g(x) \in F[x]$. Then K_1K_2 is the splitting field of $f(x)\,g(x) \in F[x]$. Hence K_1K_2/F is normal and hence it is a Galois extension. The map $G(K_1K_2/F) \to G(K_1/F) \times G(K_2/F)$ defined by $\sigma \to (\sigma \mid K_1, \sigma \mid K_2)$ is a homomorphism with kernel fixing K_1K_2. Thus the kernel is *Id*. Hence the map is an isomorphism "into".

6. Let K/F be a Galois extension and L/F any finite extension. Show that $[KL : L]$ divides $[K : F]$. Illustrate by an example that this is no longer true if K/F is not a Galois extension.

Solution: Since K/F is a Galoris extension, KL/L is a Galois extension. Let $G = G(K/F)$ and $G' = G(KL/L)$. The map $G' \to G$ given by $\sigma \to \sigma \mid K$ is an injective homomorphism.

Hence $o(G') \mid o(G)$. *i.e.* $[KL : L] \mid [K : F]$.

If the extension K/F is not a Galois extension, the result is not true.

For example, let $K = Q(w\sqrt[3]{2})$ and $F = Q$, where w is a cube root of unity.

K/F is not a Galois extension and $[K : F] = 3$. Let $L = Q(\sqrt[3]{2})$.

Then $KL = Q(\sqrt[3]{2}, w)$ and has degree 2 over L and does not divide $[K : F]$.

7. Find all automorphisms of the following field extensions and also the fixed fields in each case.

(*i*) $Q(i, \sqrt[4]{2})/Q$

(*ii*) $F(t)/F(t^2)$. F any field, and t transcendental.

Solution: (*i*) $K = Q(i, \sqrt[4]{2})/Q$ is a Galois extension of degree 8 over Q, being the splitting fields of $x^4 - 2 \in Q[x]$. From Problem 1, G is generated by $\{\sigma, \tau\}$ with relations $\sigma^4 = 1 = \tau^2$ and $\tau\sigma = \sigma^3\tau$. The fixed field in this case is Q.

(*ii*) $F(t)/F(t^2)$ is of degree 2 and hence normal and also separable. Hence the Galois group is order 2. Write any element of $F(t)$ as

$$f(t) + tg(t), f(t), g(t) \in (t^2).$$

The automorphisms are given by

$$f + tg \to f + tg \quad \text{and} \quad f + tg \to f - tg.$$

The fixed field is $F(t^2)$.

8. Let $K = C(t)$, the field of rational functions in t over C and ω a cube root of unity. Show that $\sigma(t) = \omega t$, $\sigma \mid C = Id$, and $\tau(t) = t^{-1}$, $\tau \mid C = Id$, are C-automorphisms of K satisfying $\sigma^3 = I = \tau^2$, and $\tau\sigma = \sigma^{-1}\tau$. Show that the group G generated by σ, and τ is isomorphism to S_3 and the fixed field of G is $C(\alpha)$, where $\alpha = t^3 + t^{-3}$.

Solution: Clearly σ, τ are automorphisms satisfying the specified relations.

The map $\theta : G \to S_3$ given by $\theta(\sigma) = \phi$, $\sigma(\tau) = \psi$

is an isomorphism of G "onto" S_3. Clearly σ and τ fix α and $C(\alpha) \subset$ Fixed field of G.

Since $[C(t) : C(\alpha)] = 6 = o(G)$, $C(\alpha)$ coincides with the fixed field.

9. Let $K = Z_p(t)$ be the rational function field in t over Z_p and $\sigma : K \to K$ the Z_p-automorphism of K defined by $\sigma(t) = t + 1$. Find the fixed field of $G = (\sigma)$, the cyclic group generated by σ.

Solution: Clearly σ is an automorphism of order p and by Artin's theorem, if F is the fixed field, K/F is of degree p. $t^p - t \in F$ and $Z_p(t^p - t) \subset F$ and since K has degree p over $Z_p(t^p - t)$, $Z(t^p - t) = F$.

10. Let K/F be a finite Galois extension with $G = G(K/F)$. Assume that $G = G_1 \times G_2$. Let K_1 be the fixed field of $G_1 \times (e)$ and K_2 the fixed field of $\{e\} \times G_2$. Show that K_i/F $(i = 1, 2)$ are Galois extensions, with $K_1 \cap K_2 = F$, and $K_1K_2 = K$.

Solution: Since G_i $(i = 1, 2)$ is normal in G, by fundamental theorem both K_1/F and K_2/F are Galois extensions. Every element of $K_1 \cap K_2$ is fixed both by $G_1 \times \{e\}$ and $\{e\} \times G_2$ and hence by $G_1 \times G_2 = G$. Thus $K_1 \cap K_2 = F \cdot G$ (K/K_1K_2) is a Galois extension with Galois group contained in G $(K/K_1) = G_1 \times \{e\}$ as well as G $(K/K_2) = \{e\} \times G_2$ and hence in $\{e\} \times \{e\}$. Hence G (K/K_1K_2) is trivial, *i.e.* $K = K_1K_2$.

11. Let F be any field and $K = F(x)$ the field of rational functions in x over F. Show that every F-automorphism of K is generated by automorphism of the type $x \to \dfrac{ax+b}{cx+d}$, $ad - bc \neq 0$. Show that the group of all F-automorphisms of K is generated by automorphism of the type $x \to x + a$, $x \to bx$, $b \neq 0$. Show that the order of the group is $q^3 - q$ if F is a finite field with q elements.

Solution: Let $\sigma(x) = u$. Since σ is an F-automorphism, $F(x) = F(u)$. Write $u = \dfrac{f(x)}{g(x)}$ and call maximum of degrees of $f(x)$ and $g(x)$ as degree of u.

Now x is algebraic over $F(u)$ with minimum polynomial

$$g(t)\, u - f(t) \in F(u)[t]$$

and since $F(x) = F(u)$ this must be of degree one in t. Thus $\deg u = 1$.

Hence $u = \dfrac{ax+b}{cx+d}$, and σ is an automorphism if $ad - bc \neq 0$. Clearly $x \to x + a$, $x \to bx$, $b \neq 0$ are automorphisms and they generate all these automorphisms. The number of such automorphism is number of non-singular 2×2 matrices over F_q modulo a non-zero scalar multiple.

To count the number of non-singular 2×2 matrices over F_q we have to count the number of linear maps of F_q^2 onto F_q^2. Such a map is determined by its value on two basis elements ρ_1 and ρ_2. The value on ρ_1, can be any one of the $q^2 - 1$ non-zero values and on ρ_2 the value can be any value different from a scalar multiple of the value on ρ_1 *i.e.* $q^2 - q$ in number. Hence the total number of such maps is $(q^2 - 1)(q^2 - q)$. The required number is therefore.

$$\frac{(q^2-1)(q^2-q)}{q-1} = q(q^2 - 1) = q^3 - q.$$

4.6 CYCLIC EXTENSIONS

1. Let K/F be a cyclic extension and L a field such that $F \subset L \subset K$. Show that K/L is cyclic and if L/F is normal, L/F is cyclic.

Solution: Let $G = G(K/F)$. Then G is a cyclic group. K/L is a Galois extension with $G(K/L)$ a subgroup of G, and hence cyclic. Hence K/L is a cyclic extension. If L/F is normal, it is also a Galois extension with $G(L/F)$ a quotient group of G and hence cyclic. Thus L/F is also a cyclic extension.

2. Let K/F be a cyclic extension. Show that there is a unique field L such that $F \subset L \subset K$ such that $[L : F] = d$, for each divisor d of $n = [K : F]$.

Solution: $G = G(K/F)$ is a cyclic group of order n. For each $d \mid n$, it has a unique subgroup H of order $d' = d/n$. Let L be the fixed field of H. Then $[K : L] = d'$ and hence $[L : F] = d$.

3. Let ρ be a primitive 5th root of unity in C and $K = Q(\rho)$. Verify the fundamental theorem of Galois theory for K/Q.

Solution: $[K : Q] = 4$ with minimum polynomial of ρ given by $x^4 + x^3 + x^2 + x + 1$. $G(K/Q)$ is isomorphic to the group Z_5^* which is cyclic of order 4 under the map $G(K/Q) \to Z_5^*$, $\sigma \to \bar{i}$, where $\sigma(\rho) = \rho^i$. Subgroups of Z_5^* are $\{\bar{1}\}$, Z_5^* and a unique cyclic subgroup of order 2. The corresponding fixed fields are K, Q and $Q(\rho + \rho^4)$ as $\rho \to \rho^4$ is element of order 2 in $G(K/Q)$.

4. Determine which roots of unity lie in the following fields:

(*i*) $Q(\sqrt{-2})$ (*ii*) $Q(\sqrt{-5})$.

Solution: In both cases the only roots of unity lying in these fields are ± 1. For if they contain an nth root of unity, $n > 2$, then they contain a pth root of unity for some odd prime p. The degree of such a pth root of unity over Q is $p - 1$ and since the extensions are quadratic, $p - 1$, must divide 2, *i.e.*, $p - 1 = 2$, $p = 3$. However, it is clear that the cube roots of unity cannot lie in these fields. Thus ± 1 are the only roots of unity in (*i*) as well as (*ii*).

5. For what integers n, does a primitive nth root of unity have degree 2 over Q.

Solution: Let ρ be a primitive nth root of unity. Then $\deg(\rho) = \phi(n) = 2$. Then $n = 3$.

6. Let ρ be a primitive nth root of unity in a field F, n being an odd integer. Show that F also contains primitive $2n$th roots of unity.

Solution: The $2n$th roots of unity are the roots of

$$x^{2n} - 1 = (x^n - 1)(x^n + 1).$$

F contains roots of $x^n - 1$ generated be ρ. Roots of $x^n + 1$ are $-1, -\rho, -\rho^2, \ldots, -\rho^{n-1}$ (n odd) and all these lie in F. Hence F contains primitive $2n$th roots of unity.

7. Let K be a finite extension of Q. Show that K contains only finitely many roots of unity.

Solution: Let $[K : Q] = n$. If K contains an mth root of unity, it contains a primitive dth root of unity for $d \mid m$. It is sufficient to show that K contains only finitely many primitive roots of unity. For any primitive dth root of unity in K, $\varphi(d) \mid n$, and the number of such d' s is finite, as the number of divisors of n is finite.

8. Let $\rho_1, \rho_2, ..., \rho_{\phi(n)}$ be all the primitive nth roots of unity in C/Q and let $\Phi_n(x) = \prod_{i=1}^{\phi(n)} (x - \rho_i)$. Then $\Phi_n(x)$ is called the nth *cyclotomic polynomial.* Show that

(*i*) $\Phi_n(x) \in Q(x)$.

(*ii*) Calculate $\Phi_n(x)$ explicitly for $n = 3, 4$ and show that it has integral coefficients.

(*iii*) Show that $\Phi_n(x) \in Z[x]$ for all x.

Solution: (*i*) Since σ permutes primitive nth roots of unity, σ fixes the polynomial $\Phi_n(x)$ for each $\sigma \in G\left(\frac{Q(\rho)}{Q}\right)$, ρ primite nth root of unity. Hence $\Phi_n(x)$ has coefficients in Q.

(*ii*) Using the next problem

$$\Phi_3(x) = \frac{x^3 - 1}{\Phi_1(x)} = \frac{x^3 - 1}{x - 1} = x^2 + x + 1$$

$$\Phi_4(x) = \frac{x^4 - 1}{\Phi_1(x)\,\Phi_2(x)} = \frac{x^4 - 1}{(x-1)(x+1)} = x^2 + 1$$

(*iii*) The proof is by induction on n. Assume the result for all $m < n$. Write

$$x^n - 1 = \Phi_n(x) \prod_{\substack{d|n \\ d<n}} \Phi_d(x) = \Phi_n(x) f(x)$$

By induction $f(x) \in Z[x]$. Now the quotient of dividing $x^n - 1$ by $f(x)$ is $\Phi_n(x)$ with zero remainder. Since both $x^n - 1$ and $f(x)$ are in $Z[x]$, the quotient $\Phi_n(x)$ is also in $Z[x]$.

9. Show that $x^n - 1 = \prod_{d|n} \Phi_d(x)$.

Solution: Any nth root of unity is a primitive dth root of unity for some $d|n$ and any primitive dth root of unity for $d \mid n$ is an nth root of unity. The result follows.

10. Show that $\Phi_p(x)$ is an irreducible polynomial for a prime p.

Solution: $\Phi_p(x) = x^{p-1} + x^{p-2} + ... + x + 1$

and $\Phi(x+1)$ is irreducible by Eisenstein's criterion. Hence $\Phi_p(x)$ is irreducible.

11. Show that if n is odd and $n > 1$,

$$\Phi_{2n}(x) = \Phi_n(-x)$$

Solution: The proof is by induction on n. Assume that for all odd, $d' < n$, $\Phi_{2d'}(x) = \Phi_{d'}(-x)$ except that when $d' = 1$. For

$$d' = 1,\ \Phi_{2d'}(x) = \Phi_2(x) = x + 1 = -(-x - 1) = -\Phi_1(-x)$$

Consider the relation $\quad x^{2n} - 1 = -(x^n - 1)((-x)^n - 1)$.

$$\Phi_{2n}(x) = \frac{x^{2n}-1}{\prod\limits_{\substack{d|2n \\ d<2n}} \Phi_d(x)} = \frac{-(x^n-1)((-x)^n-1)}{\prod\limits_{d|n}\Phi_d(x) \prod\limits_{\substack{d'|n \\ d'<n}} \Phi_{2d'}(x)}$$

$$= \frac{(-x)^n-1}{\prod\limits_{\substack{d'|n \\ d'<n}} \Phi_{2d'}(x)} \; as \; x^n - 1 = \prod_{d|n} \Phi_d(x)$$

$$= \frac{(-x)^n-1}{\prod\limits_{\substack{d'|n \\ 1<d'<n}} \Phi_{d'}(-x)\,\Phi_2(x)}$$

$$= \frac{(-x)^n-1}{\prod\limits_{\substack{d'|n \\ d'<n}} \Phi_{d'}(-x)} = \Phi_n(-x), \text{ since } \Phi_2(x) = -\Phi_1(-x).$$

4.7 SOLVABILITY BY RADICALS

1. Prove that a symmetric polynomial in $x_1, x_2, ..., x_n$ is a polynomial in elementary symmetric functions $a_1, ..., a_n$.

Solution: We can assume without loss of generality that the polynomial F is homogeneous in $x_1, ..., x_n$ as any polynomial is a sum of homogeneous polynomials. Let the highest monomial in f under lexicographic ordering be $x_1^{k_1}\, x_2^{k_2}\, ...\, x_n^{k_n}$.

Since the highest monomial in $a_1^{d_1}\, a_2^{d_2}\, ...\, a_n^{d_n}$ is

$$x_1^{d_1+...+d_n}\, x_2^{d_2+...+d_n}\, ...\, x_n^{d_n}$$

The highest monomial in $a_1^{k_1-k_2}\, a_2^{k_2-k_3}\, ...\, a_{n-1}^{k_{n-1}-k_n}\, a_n^{k_n}$ is $x_1^{k_1} ... x_n^{k_n}$ which is the same as the highest monomial in f.

When we subtract from f a scalar multiple of $a_1^{k_1-k_2}\, a_2^{k_2-k_3}\, ...\, a_n^{k_n}$, we get a polynomial whose highest monomial has degree less than the degree of the highest monomial of f. Repeated application of this procedure expresses f as a polynomial in $a_1, ... a_n$.

2. Express the following as polynomials in a_1, a_2, a_3.

(*i*) $x_1^2 + x_2^2 + x_3^2$;

(*ii*) $(x_1 - x_2)^2 (x_2 - x_3)^2 (x_3 - x_1)^2$.

Solution: (*i*) $x_1^2 + x_2^2 + x_3^2 = a_1^2 - 2a_2$

(*ii*) Let $\Delta = \prod\limits_{1 \le i < j \le 3} (x_i - x_j)$.

$$\text{Since} \quad \begin{vmatrix} 1 & 1 & 1 \\ x_1 & x_2 & x_3 \\ x_1^2 & x_2^2 & x_3^2 \end{vmatrix} = \prod_{i>j} (x_i - x_j)$$

by multiplying by transpose of the matrix and taking determinant

$$\Delta^2 = \begin{vmatrix} 3 & s_1 & s_2 \\ s_1 & s_2 & s_3 \\ s_2 & s_3 & s_4 \end{vmatrix}$$

where $\quad s_1 = x_1 + x_2 + x_3,\ s_2 = x_1^2 + x_2^2 + x_3^2,$

$$s_3 = x_1^3 + x_2^3 + x_3^3 \text{ and } s_4 = x_1^4 + x_2^4 + x_3^4.$$

By computation $\quad s_1 = a_1,\ s_2 = a_1^2 - 2a_2\ s_3 = a_1^3 - 3a_1a_2 + 3a_3,$

$$s_4 = a_1^4 - 4a_1^2\ a_2 + 4a_1\ a_3 + 2a_2^2.$$

Hence Δ^2 after computation and substituting for the values of s_i equals

$$-4a_1^3\ a_3 + a_1^2a_2^2 + 18a_1a_2a_3 - 4a_2^3 - 27a_3^2.$$

3. What is the splitting field of $x^2 + x + 1 \in Z_2[x]$? Is it a radical extension of Z_2? Why?

   Solution: $x^2 + x + 1$ is irreducible in $Z_2[x]$. If α is a root of this polynomial, splitting field is $Z_2(\alpha)$. It is not a radical extension of Z_2.

4. Let ch $F = 0$. Show that the polynomial

$$f(x) = ax^8 + bx^6 + cx^4 + dx^2 + e \in F[x]$$

   is solvable by radicals.

   Solution: $f(x) = g(x^2)$, where $g(x) = ax^4 + bx^3 + cx^2 + dx + e$. Now $g(x)$ is solvable by radicals, *i.e.*, roots of $g(x)$ are contained in a radical extension of F. Roots of $f(x)$ are square root of $g(x)$ and hence contained in a radical extension of F. Hence $f(x)$ is solvable by radicals.

5. Determine the Galois group of $x^4 - 3x^3 + 3x + 2$ over Q.

   Solution: $x = 2$ is a root of the polynomial so that the splitting field is the same as the splitting field of $x^3 - x^2 - 2x - 1$. We compute the discriminant of this cubic.

   The discriminant of $x^3 - a_1x^2 + a_2x - a_3$

$$\Delta = -4\ a_1^3a_3 + a_1^2a_2^2 + 18a_1a_2a_3 - 4a_2^3 - 27a_3^2.$$

   In this case $\Delta = -31$ which is not a square. Using the fact that for an irreducible cubic the group is A_3 or S_3 according as whether Δ is a square or not, it follows that the Galois group of the polynomial is S_3. (See Problem 7 below)

6. Let ch $F \neq 2$. Show that the Galois group of $x^2 - a \in F[x]$ is of order 2 if a is not a square in F.

   Solution: The polynomial $x^2 - a$ is irreducible in $F[x]$ if a is not a square and if α is a root of this polynomial, the splitting field is $F(\alpha)$ and $[F(\alpha) : F] = 2$. Hence $G(F(\alpha)/F)$ is of order 2.

7. Let ch $F \neq 2, 3$ and $f(x) = x^3 + bx + c$. Show that the Galois group of $f(x)$ is S_3 if, and only if, $\Delta = -4b^3 - 27c^2$ is not a perfect square in F.

Solution : The discriminant $\Delta = \prod_{1 \le i < j \le 3} (\alpha_i - \alpha_j)^2$

$$\begin{vmatrix} 1 & 1 & 1 \\ \alpha_1 & \alpha_2 & \alpha_3 \\ \alpha_1^2 & \alpha_2^2 & \alpha_3^2 \end{vmatrix} = \prod_{i>j} (\alpha_i - \alpha_j)$$

Multiply by transpose, $\begin{vmatrix} 3 & s_1 & s_2 \\ s_1 & s_2 & s_3 \\ s_2 & s_3 & s_4 \end{vmatrix} = \prod_{i>j} (\alpha_i - \alpha_j)^2 = \Delta$

where $s_i = \alpha_1^i + \alpha_2^i + \alpha_3^i \quad 1 \le i \le 4$

So $s_1 = a_1, s_2 = a_1^2 - 2a_2$

$$s_3 = a_1 (\alpha_1^2 + \alpha_2^2 + \alpha_3^2) - a_2 (\alpha_1 + \alpha_2 + \alpha_3) + 3a_3$$

$$= a_1^3 - 3a_1a_2 + 3a_3$$

Also $s_4 = a_2s_3 - a_2 s_2 + a_3s_1$

$$= a_1^4 - 4 a_1^2 a_2 + 4a_1a_3 + 2a_2^2 .$$

In our case $a_1 = 0, a_2 = + b, a_3 = - c$ and by computation $\Delta = - 4b^3 - 27c^2$. The action of an element σ of S_3 on Δ is given by $\sigma(\Delta) = \Delta$ if σ is even or $\sigma(\Delta) = -\Delta$, if σ is odd.

In particular if $d = \Delta^2$, $\sigma(d) = d$, for all σ. So the Galois group of $f(x)$ is a subgroup of A_3 if, and only if, Δ is a square. Hence the Galois group of $f(x)$ is S_3 if, and only if, Δ is not a square.

8. Show that any finite group be realised as the Galois group of a suitable Galois extension.

Solution: Let G be a finite group. Then G is a subgroup of S_n, where $n = o(G)$. By Abel's theorem, S_n is the Galois group of $F(x_1, ..., x_n)$ over $F(a_1, ..., a_n)$. If L is the fixed field of G, then $F(x_1, ..., x_n)$ L has Galois group equal to G

9. Let $x^p - a \in Q[x]$ be irreducible, p prime. Show that its group over Q is isomorphic to the group of transformations of Z_p of the type $x \to \lambda x + \mu$, $\lambda, \mu \in Z_p$, $\lambda \neq 0$.

Solution: The splitting field of $x^p - a$ is $Q(\rho, \alpha)$, where α is a root of $x^p - a$ in some extension field and ρ is a primitive pth root of unity.

If $K = Q(\rho, \alpha)$, $[K : Q] = p^2 - p$ and $G(K/Q)$ is of order $p^2 - p$. If $\sigma \in G(K/Q)$, σ is determined by its values on ρ and α.

If $\sigma(\alpha) = \rho^i \alpha$, $0 \le i \le p - 1$ and $\sigma(\rho) = \rho^j$; $1 \le j \le p - 1$, the pair of integers (i, j) determine σ-completely. This corresponds to a linear map $Z_p \to Z_p$ given by $x \to jx + i$, $1 \le j \le p - 1, 0 \le i \le p - 1$. This correspondence gives the required isomorphism.

10. Let $f(x) \in F[x]$ be irreducible of prime degree over F (ch $F = 0$). Let K be a splitting field of $f(x)$. Show that $f(x)$ is solvable by radicals if and only if $K = F(\alpha, \beta)$ for any two roots α and β of $f(x)$ (Hard).

Solution: Let K be a splitting field of $f(x) \in F[x]$ and $G = G(K/F)$. Then G is a group of permutations on the roots of $f(x)$, *i.e.*, on p symbols which can be taken as $0, 1, 2, \ldots, p-1$. G is also transitive, *i.e.*, for any two symbols i and j, there is an element $\sigma \in G$ with $\sigma(i) = i$. This is because $f(x)$ is irreducible.

Let L be the affine group of all transformations from $Z_p \to Z_p$ which are of the type $x \to ax + b$, $a \in Z_p^*$, $b \in Z_p$. This includes all translations H, *i.e.*, maps of the type $x \to x + b$. These are the only elements of L without fixed points. Each element of L can be identified as an element of S_p. Thus L can be identified as a subgroup of S_p and elements of H, having no fixed points correspond to p-cycles.

We prove the following Theorem:

Theorem. Let $G \subset S_p$. The G is transitive and solvable if and only if G is a subgroup of L containing H as a normal subgroup.

We first prove the following Lemma.

Lemma 1. Let $H \neq \{e\}$ be a normal subgroup of a transitive group $G \subset S_p$. Then H is also transitive.

Proof. All the H-orbits of $\{0, 1, 2, \ldots p-1\}$ have the same cardinality. Since p is prime, there is only one H-orbit. Thus H is transitive.

Proof of Theorem. Assume G is transitive and solvable. The proof is by induction on the length of a solvable series of G. Consider a solvable series of G, *viz.*,

$$G \supset G^{(1)} \supset G^{(2)} \supset \ldots \supset \{e\}$$

with abelian quotients. $G^{(1)}$ is also solvable and transitive by Lemma 1. By induction, $G^{(1)}$ is a subgroup of L containing H as a normal subgroup. We claim that G also has the same property. Clearly G contains H.

Let $\eta \in G$ and $\tau \in H$ such that $\tau(x) = x + 1$. Then $\eta\tau\eta^{-1}$ has no fixed points and so $\eta\tau\eta^{-1} \in H$ *i.e.*, $\eta\tau\eta^{-1} = x + k$ for some $k \neq 0$. Replacing x by ηx we have $\eta\tau(x) = \eta(x) + k$ *i.e.*, $\eta(x+1) = \eta(x) + k$. Repeated use of this relation gives $\eta(x) = kx + b$, where $b = \eta(0)$. Thus $\eta \in L$. So G is a subgroup of L containing H.

Conversely assume that G is a subgroup of L containing H as a normal subgroup. The map $L \to Z_p^*$ given by $\{x \to ax + b\} \to a$ is a group homomorphism which is "onto" with kernel H. Thus L contains H as a normal subgroup with quotient isomorphic to Z_p^*. Since both H and Z_p^* are solvable (being abelian) L is solvable. Since G is a subgroups of L containing H as a normal subgroup, G/H is a subgroup of Z_p^* and hence salvable. Thus G is also solvable.

We now prove the following Lemma:

Lemma 2. Let K be a splitting field of $f(x) \in F[x]$ and $G = G(K/F)$. Then $K = F(\alpha, \beta)$ for any two roots α and β of $f(x)$ if and only if G has the property that whenever $\sigma \in G$ with $\sigma(i) = i$, $\sigma(j) = j$, $i \neq j$, $\sigma = Id$.

Proof. Let $\sigma \in G = G(K/F)$ fixing two distinct elements, *i.e.*, two distinct roots α and β of $f(x)$. By assumption $K = F(\alpha, \beta)$ and so σ fixes K, *i.e.*, $\sigma = Id$.

Conversely assure that σ has the given property. Let α and β be any two roots of $f(x)$. Consider the extension $K/F(\alpha, \beta)$. This is also a Galois extension. Let $\sigma \in G(K/F(\alpha, \beta))$. Then $\sigma \in G$ and σ fixes α and β. By assumption $\sigma = Id$. So $G(K/F(\alpha, \beta))$ is trivial, *i.e.*, $K = F(\alpha, \beta)$.

We now complete the solution to the problem. Assume that $f(x)$ is solvable by radicals. Then G is solvable. By Theorem, G is a subgroup of L containing H as a normal subgroup. We claim that G has the property in Lemma 2. Let $\sigma \in G$, with $\sigma(i) = i$, $\sigma(j) = j$, $i \neq j$. Write $\sigma(x) = ax + b$. Then $i = ai + b$ and $j = aj + b$ and so $i - j = a(i - j)$. Thus $a = 1$ and $\sigma(x) = x + b$. Now $i = i + b$ gives $b = 0$, so that $\sigma(x) = x$, $\sigma = Id$.

By Lemma 2, $K = F(\alpha, \beta)$ for any two roots α and β of $f(x)$.

For the converse, assume now that $K = F(\alpha, \beta)$ for any two roots α and β. We show that G is solvable . By Lemma 2, G has the property that for $\sigma \in G$, if $\sigma(i) = i$, $\sigma(j) = j$, $i \neq j$ then $\sigma = Id$. Since $\deg f(x) = p$, p divides the order of G and by Cauchy's theorem G has an element of order p. So G contains a p-cycle σ. We claim that $H = (\sigma)$ is normal subgroup of G. Let $\tau \in G$. Then $t\sigma t^{-1}$ is also a p-cycle and so far each i, there exists some $n(i)$ such that $\tau\sigma\tau^{-1}\sigma^{-n(i)}(i) = i$, $0 \leq i \leq p - 1$. Now $n(i) \neq 0$ as $\tau\sigma\tau^{-1}$ has no fixed point. Consider the map $i \to n(i)$ from the set $\{0, 1, \ldots, p-1\}$ to $\{1, 2, \ldots, p-1\}$. This map cannot be injective. Hence for some $i \neq j$, $n(i) = n(j) = n$ say.

Then $\tau\sigma\tau^{-1}\sigma^{-n}(i) = i$, and $\tau\sigma\tau^{-1}\sigma^{-n}(j) = j$, $i \neq j$ and so $t\sigma t^{-1}\sigma^{-n} = Id$, *i.e.*, $\tau\sigma\tau^{-1} = \sigma^n \in H$. Thus H is a normal subgroup of G.

By the same argument as in theorem, G is a subgroup of L, for if $\eta \in G$, then $\eta(x) = kx + \eta(0)$. Then G is a subgroup of L containing H as a normal subgroup. Hence by theorem, G is solvable and so $f(x)$ is solvable by radicals.

4.8 GEOMETRIC CONSTRUCTIONS

1. Starting with points of Q^2 and repeating the ruler and compass constructions a finite number of times, show that the points obtained are precisely of the type (α, β) where α and β are constructible real numbers.

Solution: The ruler and compass constructions involve (*i*) intersection of two lines, (*ii*) intersection of a line with a circle and (*iii*) intersection of two circles. The last can be reduced to the intersection of a circle with a line, namely the common chord of the two circles. Equation of a line is of the type $ax + by + c = 0$, and equation of a circle is

$$x^2 + y^2 + 2gx + 2fy + c = 0.$$

Simultaneous solution of two linear equations with rational coefficients leads only to points with rational coefficients. Solving a linear equation and a quadratic equation with rational coefficients yields points which are in a quadratic extension of Q. Repetition of these operations, a finite number of times, yields points of the type (α, β) where both α, β belong to a radical extension which is a tower of simple quadratic radical extensions. Hence they are constructible.

2. Prove that a regular pentagon is constructible.

Solution: We have to show that $\alpha = \cos 2\pi/5$ is a constructible number. Let ξ be a primitive 5th root of unity and K splitting field of $x^5 - 1 \in Q[x]$. Then $[K : Q] = \phi(5) = 4$. The subgroup of $G(K/Q)$ fixing $\xi + 1/\xi$ is of order 2 and hence $[Q(\xi + 1/\xi) : Q] = 2$; a power of 2.

Hence $\xi + 1/\xi = 2 \cos 2\pi/5$ is constructible.

3. Show that it is impossible to trisect an angle equal to 72° with ruler and compass.

Solution: If 72° can be trisected, a segment of length $\alpha = \cos 24°$ would be constructible. If $\theta = 24°$, $\cos 3\theta = 4\cos^3\theta - 3\cos\theta$ so that $\eta = \cos 72° = 4\alpha^3 - 3\alpha$, will also be constructible. α satisfies $4\alpha^3 - 3\alpha - \eta = 0$ and deg α over $Q(\eta)$ is 3. Since deg η over Q is 2, deg α over Q is not a power of 2. Hence α is not constructible.

4. Investigate whether a regular hexagon is constructible.

Solution: If a regular hexagon is constructible, $\alpha = \cos 2\pi/6$ will be constructible. Let $x^6 - 1 \in Q[x]$ and K its splitting field over Q. Then $K = Q(\xi)$, where ξ is a primitive 6th root of unity and $G(K/Q)$ is a group of order $\varphi(6) = 2$. We have $2\alpha = (\xi + 1/\xi)$ and the subgroup of $G(K/Q)$ fixing $\xi + 1/\xi$ is of order 2 so that $[Q(\xi + 1/\xi) : Q] = 2$, a power of 2. Hence $\xi + 1/\xi = 2 \cos \pi/6$ is constructible, *i.e.*, a regular hexagon is constructible.

5. Find the smallest angle in terms of degrees which is constructible by ruler and compass.

Solution: Constructing 1° by ruler and compass amounts to constructing a regular polygon of 360 sides. A regular polygon of n sides is constructible if, and only if, $\phi(n)$ is a power of 2. $\phi(360) = \phi(2^3 \cdot 3^2 \cdot 5) = 2^2 \cdot 3 \cdot 4$ is not a power of 2. Similarly if $\theta = 2°$, $\phi(180)$ is not a power of 2. But if $\theta = 3°$, $Q(120) = \phi(2^3 \cdot 3 \cdot 5)$ is a power of 2. Hence $\theta = 3°$ is constructible.

4.9 TRANSCENDENTAL EXTENSIONS

1. Let K/F be an extension of finite transcendence degree and L an extension such that $F \subset L \subset K$. Show that Tr. $\deg_F L < \infty$.

Solution: Suppose Tr : $\deg_F L = \infty$. Then there exists an infinite sequence of elements of L, $\alpha_1, \alpha_2, \ldots\alpha_n, \ldots$ which are algebraically independent over F. These are also elements of K, algebraically independent over F. Hence Tr. $\deg_F K = \infty$, a contradiction to the assumption.

2. Let $F \subset K \subset L$ be extension fields. Show that

$$\text{Tr. deg}_F L = \text{Tr. deg}_F K + \text{Tr. deg}_F K.$$

Solution: Let $\{x_1, \ldots x_n\}$ be a transcendence base of K/F and $\{y_1 \ldots y_m\}$ a transcendence base of L/K. Then $\{x_1, \ldots, x_n, y_1, \ldots, y_m\}$ is algebraically independent over F, for if there is an F algebraic relation between them, it must involve some y_i and it would mean $\{y_1 \dots y_m\}$ is algebraically dependent over $F(x_1, \dots, x_n) \subset K$, a contradiction.

Now $K/F\ (x_1, \dots x_n)$ is algebraic and $L/K\ (y_1 \dots y_m)$ is algebraic. This implies that $\dfrac{K(y_1 \dots y_m)}{F(x_1 \dots x_n, y_1 \dots y_m)}$ is algebraic and since $\dfrac{L}{K(y_1 \dots y_m)}$ is algebraic, we have $\dfrac{L}{F(x_1 \dots x_n, y_1 \dots y_m)}$ is algebraic.

Hence Tr. $\deg_F L = n + m =$ Tr. $\deg_F K +$ Tr. $\deg_K L$.

3. Let K_1/F and K_2/F be f.g. extensions, both contained in some field K. Show that

$$\text{Tr. deg } K_1K_2/F \le \text{Tr. deg}_F\, K_1 + \text{Tr. deg}_F\, K_2.$$

Solution: Let $\{x_1 \dots x_n\}$ be transcendence base of K_1 over F and $\{y_1 \dots y_m\}$ be a transcendence base of K_2 over F. Then $K_1/F\ (x_1 \dots x_n)$ is algebraic and $K_2/F\ (y_1 \dots y_m)$ is algebraic.

Then $K_1(y_1 \dots y_m)$ is algebraic over $F(x_1, \dots x_n, y_1, y_2, \dots, y_m)$. Since K_1K_2 is algebraic over $K_1(y_1 \dots y_m)$ it is also algebraic over $F(x_1 \dots x_n, y_1 \dots y_m)$.

Choose a maximal algebraically independent subset of $\{x_1 \dots x_n, y_1 \dots y_m\}$. It is a transcendence base of $K_1\, K_2$ over F. Then Tr. $\deg_F K_1K_2 \le m + n =$ Tr. $\deg_F K_1$ + Tr. $\deg_F K_2$.

CHAPTER 5

Matrices and Linear Transformations

5.1 MATRICES

1. Find AB if A and B in $M_4(2)$ are given by

$$A = \begin{pmatrix} 1 & 0 & -1 & 0 \\ 2 & 3 & 4 & 8 \\ 3 & 1 & 6 & 9 \\ 0 & 0 & 1 & 3 \end{pmatrix} \text{ and } B = \begin{pmatrix} 0 & -1 & -1 & 0 \\ 2 & 4 & 6 & 9 \\ -7 & 8 & 0 & 3 \\ 3 & 1 & 6 & 4 \end{pmatrix}$$

Solution:
$$AB = \begin{pmatrix} 7 & -9 & -1 & -3 \\ 2 & 50 & 64 & 71 \\ -13 & 58 & 57 & 63 \\ 2 & 11 & 18 & 15 \end{pmatrix}.$$

2. If $A \in M_3(R)$ is given by $A = \begin{pmatrix} 1 & 3 & 0 \\ -1 & 2 & 1 \\ 1 & 1 & 0 \end{pmatrix}$, find $A^3 + 2A^2 - A + 5I$.

Solution:
$$\begin{pmatrix} -8 & 15 & 9 \\ -2 & 0 & 2 \\ -4 & 11 & 5 \end{pmatrix} + 2\begin{pmatrix} -2 & 9 & 3 \\ -2 & 2 & 2 \\ 0 & 5 & 1 \end{pmatrix} - \begin{pmatrix} 1 & 3 & 0 \\ -1 & 2 & 1 \\ 1 & 1 & 0 \end{pmatrix} + \begin{pmatrix} 5 & 0 & 0 \\ 0 & 5 & 0 \\ 0 & 0 & 5 \end{pmatrix}$$

$$= \begin{pmatrix} -8 & 30 & 15 \\ -5 & 9 & 5 \\ -5 & 20 & 12 \end{pmatrix}.$$

3. Show that the ring $M_n(F)$ has zero divisors.

Solution:

In $M_2(\mathbb{R})$, $\begin{pmatrix} 1 & 0 \\ 0 & 0 \end{pmatrix}$ is a zero divisor because $\begin{pmatrix} 1 & 0 \\ 0 & 0 \end{pmatrix}\begin{pmatrix} 0 & 0 \\ 0 & 1 \end{pmatrix} = \begin{pmatrix} 0 & 0 \\ 0 & 0 \end{pmatrix}$.

4. For any field F, show that $M_2(F)$ has at least six non-singular elements. List all of them. (ch $F \neq 2$).

Solution: The following are the non-singular elements of $M_2(F)$:

$$\begin{pmatrix} 1 & 0 \\ 0 & 1 \end{pmatrix}\begin{pmatrix} 0 & 1 \\ 1 & 0 \end{pmatrix}\begin{pmatrix} 1 & 1 \\ 1 & 0 \end{pmatrix}\begin{pmatrix} 1 & 1 \\ 0 & 1 \end{pmatrix}\begin{pmatrix} 1 & 0 \\ 1 & 1 \end{pmatrix}\begin{pmatrix} 0 & 1 \\ 1 & 1 \end{pmatrix}.$$

5. Show that if A and B are non-singular, so is AB and $(AB)^{-1} = B^{-1} A^{-1}$.

Solution: Let A^{-1} and B^{-1} be inverses of A and B.

Then $(AB)(B^{-1}A^{-1}) = A(BB^{-1})A^{-1} = AIA^{-1} = AA^{-1} = I$

and $(B^{-1}A^{-1})(AB) = B^{-1}(A^{-1}A)B = B^{-1}IB = B^{-1}B = I$

Hence AB is non-singular and $(AB)^{-1} = B^{-1}A^{-1}$.

6. Show that a diagonal matrix with diagonal elements $d_1, d_2, \ldots, d_n$ is non-singular if, and only if, each $d_i \neq 0$. In that case, show that the inverse is the diagonal matrix with diagonal elements $d_1^{-1}, d_2^{-2}, \ldots, d_n^{-1}$.

Solution: Suppose $D = \begin{pmatrix} d_1 & & \\ & \ddots & \\ & & d_n \end{pmatrix}$ is non-singular. Then there exists a matrix $A = (a_{ij})$ such that $DA = I = AD$. Equating $DA = I$, we get $a_{11}d_1 = 1, a_{22}d_2 = 1, \ldots, a_{nn}\, d_n = 1$. Hence $d_i \neq 0$ for all i.

The product $\begin{pmatrix} d_1^{-1} & & \\ & \ddots & \\ & & d_n^{-1} \end{pmatrix}\begin{pmatrix} d_1 & & \\ & \ddots & \\ & & d_n \end{pmatrix} = \begin{pmatrix} d_1 & & \\ & \ddots & \\ & & d_n \end{pmatrix}\begin{pmatrix} d_1^{-1} & & \\ & \ddots & \\ & & d_n^{-1} \end{pmatrix} = I$. Hence D^{-1} is the diagonal matrix with diagonal elements $d_1^{-1}, \ldots, d_n^{-1}$.

7. Show that if A is a nilpotent matrix then $I + A$ is non-singular.

Solution: Let $A^m = 0$. Then $(I + A)(I - A + A^2 - A^3 + \ldots (-1)^{m-1}A^{m-1}) = I$

$$= (I - A + A^2 + \ldots + (-1)^{m-1}A^{m-1})(I + A).$$

Hence $I + A$ is non-singular and

$$(I + A)^{-1} = I - A + A^2 + \ldots + (-1)^{m-1}A^{m-1}.$$

8. Show that sum of two nilpotent matrices is nilpotent and the product of two nilpotent matrices is nilpotent if they commute with each other.

Solution: Let A and B be nilpotent so that $A^m = 0$ and $B^t = 0$.

Then $(A + \mathrm{B})^{m+1} = \sum_{i+j=m+t} A^i B^j$

since A and B commute. In the summation either $i \geq m$ or $j \geq t$ and hence each term is zero. Hence $(A + B)^{m+1} = 0$, *i.e.*, $A + B$ is nilpotent.

Now $(AB)^{mt} = A^{mt} B^{mt}$, since A and B commute. Hence $(AB)^{mt} = 0$, *i.e.* AB is nilpotent.

9. If A is an $n \times n$ triangular matrix, all of whose diagonal elements are zero, show that A is nilpotent.

Solution: If $A = \begin{pmatrix} 0 & * \dots * \\ \dots\dots\dots\dots \\ 0 & 0 \end{pmatrix}$, then $A = \begin{pmatrix} 0 & * \dots * \\ 0 & \\ \vdots & A_1 \\ 0 & \end{pmatrix}$ where A_1 is a triangular matrix with all diagonal elements zero.

By computation $A^2 = \begin{pmatrix} 0 & 0 & \dots * \\ 0 & & \\ \vdots & & A_1^2 \\ 0 & & \end{pmatrix}$, $A^3 = \begin{pmatrix} 0 & 0 & 0 & * \dots * \\ 0 & & & \\ \vdots & & & A_1^3 \\ 0 & & & \end{pmatrix}$

$A^n = \begin{pmatrix} 0 & 0 \dots 0 \\ 0 & \\ \vdots & A_1^n \\ 0 & \end{pmatrix}$. By induction on the size of A, $A_1^{n-1} = 0$ and hence $A_1^n = 0$, *i.e.* $A^n = 0$. Hence A is nilpotent. For 2×2 matrix the result is clear because

$$\begin{pmatrix} 0 & * \\ 0 & 0 \end{pmatrix}\begin{pmatrix} 0 & * \\ 0 & 0 \end{pmatrix} = \begin{pmatrix} 0 & 0 \\ 0 & 0 \end{pmatrix}.$$

10. Show that a triangular matrix is non-singular if, and only if, all the diagonal elements are non-zero.

Solution: Let A be a triangular matrix which is non-singular. Then there exists a matrix B with $AB = I$, *i.e.*

$$\begin{pmatrix} a_{11} & 0 & \dots & 0 \\ a_{12} & a_{22} & \dots & 0 \\ \dots & \dots & \dots & a_{nn} \end{pmatrix}\begin{pmatrix} b_{11} & \dots & b_{1n} \\ \dots & \dots & \dots \\ b_{n1} & \dots & b_{nn} \end{pmatrix} = \begin{pmatrix} 1 & 0 & \dots & 0 \\ & & & \\ 0 & 0 & \dots & 1 \end{pmatrix}$$

$$a_{11}b_{11} = 1, a_{22}b_{22} = 1, \dots a_{nn}b_{nn} = 1.$$

Hence $a_{ij} \neq 0$ for all i.

Conversely, assume that A is a triangular matrix with non-zero diagonal elements. Then A is non-singular by induction on n.

For $n = 2$, $A = \begin{pmatrix} a & 0 \\ b & c \end{pmatrix}$, $a, c \neq 0$. Hence $ac \neq 0$ *i.e.* A is non-singular.

Assume the result for $n - 1$. Let A be an $n \times n$ triangular matrix with non-zero diagonal elements. Write $A = \begin{pmatrix} a & 0 \dots 0 \\ * & \\ \vdots & A_1 \\ * & \end{pmatrix}$, where A_1 is an $(n - 1) \times (n - 1)$

triangular matrix with non-zero diagonal elements. Then by induction, there exists an $(n-1) \times (n-1)$ matrix B_1 such that $A_1B_1 = B_1A_1 = I_{n-1}$. Let B be the

matrix $B = \begin{pmatrix} a^{-1} & 0 \dots 0 \\ * & \\ \vdots & B_1 \\ * & \end{pmatrix}$. The elements of B in the first column below a^{-1} can be so chosen that $AB = BA = I_n$. Hence A is non-singular.

11. Determine all 2×2 matrices which are (*i*) non-singular, (*ii*) nilpotent, (*iii*) idempotent.

Solution:

(*i*) There are 6 non-singular 2×2 matrices and these are

$$\begin{pmatrix} 1 & 0 \\ 0 & 1 \end{pmatrix}\begin{pmatrix} 0 & 1 \\ 1 & 0 \end{pmatrix}\begin{pmatrix} 1 & 1 \\ 1 & 0 \end{pmatrix}\begin{pmatrix} 1 & 1 \\ 0 & 1 \end{pmatrix}\begin{pmatrix} 1 & 0 \\ 1 & 1 \end{pmatrix}\begin{pmatrix} 0 & 1 \\ 1 & 1 \end{pmatrix}.$$

(*ii*) *Nilpotent case*: Any 2×2 nilpotent matrix is similar to a triangular matrix with diagonal elements zero.

Here $A = \begin{pmatrix} 0 & 0 \\ a & 0 \end{pmatrix}$, a real, or $A = \begin{pmatrix} 0 & a \\ 0 & 0 \end{pmatrix}$.

(*iii*) *Idempotent case*: Any idempotent matrix is similar to a diagonal matrix so that A can be assumed to be of the form $\begin{pmatrix} d_1 & 0 \\ 0 & d_2 \end{pmatrix} = A.A^2 = A$ implies $d_1^2 = d_1$ and $d_2^2 = d_2$ so that $d_1 = 1$ or 0 and $d_2 = 1$ or 0. Hence the required matrices are zero matrix, identity matrix, $\begin{pmatrix} 1 & 0 \\ 0 & 0 \end{pmatrix}$ and $\begin{pmatrix} 0 & 0 \\ 0 & 1 \end{pmatrix}$.

12. Show that for any two $A, B \in M_2(F)$, the matrix $(AB - BA)^2$ is a scalar matrix.

Solution: This can be verified by direct computation.

13. Show that an $n \times n$ matrix which commutes with all the diagonal matrices is itself a diagonal matrix. Deduce that the centre of $M_n(F)$ is the set of all scalar matrices.

Solution: Let $A = (a_{ij})$ and $D = \begin{pmatrix} d_1 & & 0 \\ & \ddots & \\ 0 & & d_n \end{pmatrix}$. Equating AD and DA we have

$a_{12}(d_1 - d_2) = 0, \dots, a_{1n}(d_1 - d_n) = 0.$

We can choose all the d_i's to be distinct so that $a_{12} = 0, \dots a_{1n} = 0$. Similarly, by looking at the other entries, we get $a_{ij} = 0$ for $i \neq j$. Hence A is a diagonal

matrix. So let $A = \begin{pmatrix} a_1 & & \\ & \ddots & \\ & & a_n \end{pmatrix}$. Let $D = \begin{pmatrix} 0 & 1 & 0 & 0 & \dots & 0 \\ 0 & 0 & 0 & \dots & \dots & 0 \\ 0 & \dots & \dots & \dots & \dots & 0 \end{pmatrix}$. Equating

$AD = DA$ we have $a_1 = a_2$. Similarly by choosing other D's we get $a_1 = a_2 = \ldots$ a_n. Hence A is a scalar matrix. Thus the centre of A consists only of scalar matrices.

5.2 MATRICES AND LINEAR TRANSFORMATIONS

1. Let $V = R^3$, and let $T : V \to V$ be the linear transformation defined by $T(x, y, z) = (x', y', z')$ where $x' = 2x$, $y' = 4y$, $z' = 5z$. Find the matrix of T with respect to the basis (2/3, 0, 0), (0, 1/2, 0) and (0, 0, 1/4).

Solution:

Now $T(2/3, 0, 0) = (4/3, 0, 0) = 2\,(2/3, 0, 0)$

$T(0, 1/2, 0) = (0, 2, 0) = 4\,(0, 1/2, 0)$

$T(0, 0, 1/4) = (0, 0, 5/4) = 5\,(0, 0, 1/4)$

Hence the required matrix is $\begin{pmatrix} 2 & 0 & 0 \\ 0 & 4 & 0 \\ 0 & 0 & 5 \end{pmatrix}$.

2. Let $V = R^3$ and let $A = \begin{pmatrix} 1 & 2 & -3 \\ 3 & 1 & -5 \\ 0 & 0 & 1 \end{pmatrix}$ be the matrix of $T : V \to V$ with respect to the standard basis (1, 0, 0) (0, 1, 0) and (0, 0, 1). Find the matrix of T with respect to the basis (1, 1, 0), (0, 1, 0) and (0, 1, 1.)

Solution:

Now (1, 1, 0) = (1, 0, 0) + (0, 1, 0)

and (0, 1, 1) = (0, 1, 0) + (0, 0, 1)

So the matrix C which takes the standard basis to the new basis is

$C = \begin{pmatrix} 1 & 1 & 0 \\ 0 & 1 & 0 \\ 0 & 1 & 1 \end{pmatrix}$. The required matrix with respect to the new basis is $B = CAC^{-1}$, which can be computed.

3. Let $V = C$ be considered as a vector space over $\mathbb{R}$ and $T_c : V \to V$ the linear map which is right multiplication by c. Find the matrix $m\,(T_c)$ of this linear map with respect to the standard basis $\{1, i\}$. Show further that the correspondence $c \to m\,(T_c)$, defines as isomorphism of C onto $M_2\,(R)$.

Solution: Let $c = a + ib$. $T_c\,(1) = a + ib$ and $T_c\,(i) = ia - b$. So $m(T_c) = \begin{pmatrix} a & b \\ -b & a \end{pmatrix}$.

Consider the map $\theta : C \to M_2\mathbb{R}$

$$a + ib \to \begin{pmatrix} a & b \\ -b & a \end{pmatrix}$$

θ is a linear map and can be verified to be an isomorphism.

4. Let $V = \mathbb{R}^3$ and $\{v_1, v_2, v_3\}$ be a basis of V. Define $T : V \to V$ by $T(v_i) = v_1 + v_2 + v_3$, $1 \le i \le 3$. Expressing T with respect to another suitably chosen basis show that $\begin{pmatrix} 1 & 1 & 1 \\ 1 & 1 & 1 \\ 1 & 1 & 1 \end{pmatrix}$ is similar to $\begin{pmatrix} 3 & 0 & 0 \\ 0 & 0 & 0 \\ 0 & 0 & 0 \end{pmatrix}$.

Solution: Matrix of T with respect to $\{v_1, v_2, v_3\}$ is $\begin{pmatrix} 1 & 1 & 1 \\ 1 & 1 & 1 \\ 1 & 1 & 1 \end{pmatrix}$. Consider a new basis $\{w_1, w_2, w_3\}$, where $w_1 = v_1 + v_2 + v_3$, $w_2 = -v_1 + v_2$, $w_3 = -v_1 + v_3$. The matrix of T with respect to this basis is $\begin{pmatrix} 3 & 0 & 0 \\ 0 & 0 & 0 \\ 0 & 0 & 0 \end{pmatrix}$. Hence it is similar to $\begin{pmatrix} 1 & 1 & 1 \\ 1 & 1 & 1 \\ 1 & 1 & 1 \end{pmatrix}$.

5. Under what conditions on x, will the matrix $\begin{pmatrix} 1 & x \\ 0 & 1 \end{pmatrix}$ be similar to $\begin{pmatrix} 1 & 0 \\ 0 & 1 \end{pmatrix}$?

Solution: Any matrix similar to $\begin{pmatrix} 1 & 0 \\ 0 & 1 \end{pmatrix}$ is $\begin{pmatrix} 1 & 0 \\ 0 & 1 \end{pmatrix}$. So the condition is that $x = 0$.

6. Let $V = F^{(m)}$ and $V' = F^{(n)}$ and $A = (a_{ij}) \in M_{m,n}(F)$. Define a linear transformation $T_A : V \to V'$ by $T_A(x_1, \ldots, x_m) = (y_1, \ldots, y_n)$ where $y_i = \sum_{j=1}^{m} a_{ji}\, x_j$, $1 \le i \le n$.

Show that the matrix of T_A with respect to the standard basis of V is A, thereby realising any matrix as the matrix of a suitable linear transformation.

Solution: To compute $m(T_A)$, take its values on the basis elements.

$$T_A(e_1) = T_A(1, 0, \ldots, 0) = (a_{11}, \ldots a_{1n}) = a_{11}\, e_1 + \ldots + a_{1n}\, e_n$$
$$T_A(e_2) = T_A(0, 1, \ldots, 0) = (a_{21}, \ldots a_{2n}) = a_{21}\, e_1 + \ldots + a_{2n}$$

....

$$T_A(e_n) = T_A(0, 0, \ldots, 1) = (a_{m1} \ldots, a_{mn}) = a_{m1}\, e_1 + \ldots + a_{mn}\, e_n$$

Thus the matrix of T_A is A.

7. By direct computation, show that the matrices $A = \begin{pmatrix} 0 & 1 & 0 \\ 0 & 0 & 1 \\ 0 & 0 & 0 \end{pmatrix}$ and $B = \begin{pmatrix} 0 & 1 & 0 \\ 0 & 0 & 1 \\ 1 & 0 & 0 \end{pmatrix}$ are not similar.

Solution: Suppose they are similar, then $B = CAC^{-1}$ for some non-singular matrix $C = (C_{ij})$.

Then $BC = CA$ i.e., $\begin{pmatrix} c_{11} & c_{12} & c_{13} \\ c_{21} & c_{22} & c_{23} \\ c_{31} & c_{32} & c_{33} \end{pmatrix}\begin{pmatrix} 0 & 1 & 0 \\ 0 & 0 & 1 \\ 0 & 0 & 0 \end{pmatrix} = \begin{pmatrix} 0 & 1 & 0 \\ 0 & 0 & 1 \\ 1 & 0 & 0 \end{pmatrix}\begin{pmatrix} c_{11} & c_{12} & c_{13} \\ c_{21} & c_{22} & c_{23} \\ c_{31} & c_{32} & c_{33} \end{pmatrix}$.

Equating the entries of the matrix (first column) $c_{21} = 0$, $c_{31} = 0$, $c_{11} = 0$ and this makes C singular a contradiction.

8. If A and B are similar matrices determines whether the following pairs of matrices are similar:

(*i*) aA and aB, $a \in F$, $a \neq 0$

(*ii*) A^n and B^n

(*iii*) A^{-1} and B^{-1}, both non-singular.

Solution:

(*i*) Yes. If $B = CAC^{-1}$, $aB = C\,(aA)\,C^{-1}$.

(*ii*) Yes. $B^n = (CAC^{-1}) = CA^n C^{-1}$

(*iii*) Yes. $B^{-1} = CA^{-1} C^{-1}$.

9. Let $T, T' \in L(V, V)$ be such that the matrix of T with respect to one basis of V is the matrix of T' with respect to another basis of V. Show that there exists an isomorphism S to V "onto" V, such that $T' = STS^{-1}$.

Solution: Let $T(e_i) = \sum_i a_{ij}\, e_j$.

and $T'(f_i) = \sum_j a_{ij}\, f_j$.

where $\{e_i\}$, $\{f_i\}$ are bases.

Define $S : V \to V$ such that $S\,(e_i) = f_i$, $1 \leq i \leq n$ and extend linearly. S is an isomorphism since it carries a basis "onto" a basis.

Now $S^{-1}\,(f_i) = e_i$ and $TS^{-1}\,(f_i) = \sum_j a_{ij}\, e_j$

$$STS^{-1}\,(f_i) = \sum_j a_{ij}\, S(e_j) = \sum_j a_{ij}\, f_j = T'(f_i)$$

Since this is true for all i, $1 \leq i \leq n$, $STS^{-1} = T'$.

10. Let $V = \mathbb{R}^3$ and $T : V \to V$ a linear map such that the matrix of T with respect to the standard basis is $A = \begin{pmatrix} 1 & 0 & 1 \\ 0 & 1 & 1 \\ 1 & 1 & 0 \end{pmatrix}$. Show that T is an isomorphism of V "onto" V. Find the matrix of T^{-1} with respect to the same basis and verify that it is the inverse of A.

Solution: Suppose $T(x, y, z) = 0$.

Then $(x + z, y + z, x + y) = 0$.

This implies that $x = y = z = 0$. Hence T is (1, 1) and is an isomorphism. Let e_1, e_2, e_3 be the standard basis.

$$T(e_1) = e_1 + e_3,\ T(e_2) = e_2 + e_3 \text{ and } T(e_3) = e_1 + e_2.$$

Since $$e_1 = 1/2\ [(e_1 + e_2) + (e_1 + e_3) - (e_2 + e_3)]$$

$$T^{-1}(e_1) = \frac{1}{2}(e_3 + e_1 - e_2)$$

Similarly $$T^{-1}(e_2) = \frac{1}{2}(e_2 + e_3 - e_1)$$

and $$T^{-1}(e_3) = \frac{1}{2}(e_1 + e_2 - e_3).$$

Hence the matrix is T^{-1} is $\frac{1}{2}\begin{pmatrix} 1 & -1 & 1 \\ -1 & 1 & 1 \\ 1 & 1 & -1 \end{pmatrix}$. By multiplying this matrix with A we get I so that it is the inverse of A.

11. Show that the matrix $\begin{pmatrix} 1 & 1 \\ 0 & 1 \end{pmatrix}$ can never be similar to a diagonal matrix.

Solution: Suppose there exists $C = \begin{pmatrix} a & b \\ c & d \end{pmatrix}$, $ad - bc \neq 0$ such that $C\begin{pmatrix} 1 & 1 \\ 0 & 1 \end{pmatrix} C^{-1}$ $= \begin{pmatrix} d_1 & 0 \\ 0 & d_2 \end{pmatrix}$. By computation this implies that $a = 0 = c$, a contradiction as $ad - bc \neq 0$.

12. Let $\{v_1, v_2, \ldots, v_n\}$ be a basis of V over F. Find the matrices of the following linear transformations with respect to $\{v_1, v_2, \ldots, v_n\}$.

(*i*) $T : V \to V$ which interchanges v_i and v_j, $i \neq j$

(*ii*) $T : V \to V$ which maps $v_i \to v_i + av_j$, $a \in F$, $i \neq j$ and keeps all the other v_k's fixed.

(*iii*) $T : V \to V$ which maps $v_i \to av_i$, $a \in F$, $a \neq 0$ and keeps all the other v_k's fixed. Indicate how the matrices in all these cases can be obtained from I_n.

Solution:

(*i*) The matrix is
$$\begin{pmatrix} 1 & 0 & \dots & \dots & \dots & 0 \\ 0 & 1 & \dots & \dots & \dots & 0 \\ 0 & 0 & \dots & 1 & \dots & 0 \\ \dots & \dots & \dots & \dots & \dots & \dots \\ 0 & 0 & 1 & \dots & \dots & 0 \\ 0 & 0 & \dots & \dots & \dots & 1 \end{pmatrix} \begin{matrix} \\ \\ \to i\text{th row} \\ \to j\text{th row} \\ \\ \\ \end{matrix}$$

i.e., *i*th and *j*th rows are interchanged in I_n.

(*ii*) The matrix is
$$\begin{pmatrix} 1 & 0 & \dots & \dots & \dots & \dots & 0 \\ 0 & \dots & 1 & \dots & a & \dots & 0 \\ \dots & \dots & \dots & \dots & \dots & \dots & \dots \\ \dots & \dots & \dots & \dots & 1 & \dots & \dots \\ 0 & 0 & \dots & \dots & \dots & \dots & 1 \end{pmatrix} \begin{matrix} \\ \to i\text{th row} \\ \to j\text{th row} \\ \\ \\ \end{matrix}$$

i.e., *a* times the *j*th row in I_n is added to the *i*th row.

(*iii*) The matrix is
$$\begin{pmatrix} 1 & 0 & \dots & \dots & 0 \\ \dots & \dots & \dots & \dots & \dots \\ 0 & 0 & a & \dots & 0 \\ \dots & \dots & \dots & \dots & \dots \\ 0 & 0 & \dots & \dots & 1 \end{pmatrix} \begin{matrix} \\ \\ \to i\text{th row} \\ \\ \\ \end{matrix}$$

i.e. *i*th row of I_n is multiplied by *a*.

So in all cases the matrices can be obtained from I_n by either

(*i*) Interchanging two rows

(*ii*) Add multiple of one row to another row

(*iii*) Multiply a row by a non-zero scalar.

(These are called the elementary row operations on I_n.)

5.3 TRACE AND TRANSPOSE

1. Let $V = \mathbb{R}^2$ and T be the linear transformation which is rotation through angle $\theta°$ about the origin. What is the matrix A of T with respect to the standard basis (1, 0) and (0, 1)? To what linear transformation does A^t correspond?

Solution:

Since
$$x' = x\cos\theta - y\sin\theta$$
$$y' = x\sin\theta + y\cos\theta$$

The matrix of T is $A = \begin{pmatrix} \cos\theta & \sin\theta \\ -\sin\theta & \cos\theta \end{pmatrix}$

A^t is $= \begin{pmatrix} \cos\theta & -\sin\theta \\ \sin\theta & \cos\theta \end{pmatrix} = \begin{pmatrix} \cos(-\theta) & \sin(-\theta) \\ -\sin(-\theta) & \cos(-\theta) \end{pmatrix}$ corresponds to rotation through $(-\theta)$.

So A^t is the inverse of A.

2. Show that if A, B and AB are symmetric matrices, then $AB = BA$.

Solution:

$$A = A^t,\ B = B^t \text{ and } (AB) = (AB)^t$$

Now $$AB = (AB)^t = B^tA^t = BA.$$

3. Give an example of a non-singular matrix A for which $A^{-1} \neq A^t$.

Solution: Let $A = \begin{pmatrix} 1 & 2 \\ 3 & 4 \end{pmatrix}$

Then $$A^{-1} = \frac{1}{2}\begin{pmatrix} -4 & 2 \\ 3 & -1 \end{pmatrix} \text{ and } A^t = \begin{pmatrix} 1 & 3 \\ 2 & 4 \end{pmatrix}$$

Thus $A^{-1} \neq A^t$.

4. Show that for any $n \times n$ matrix, A, AA^t is symmetric.

Solution: $(AA^t)^t = (A^t)^t A^t = AA^t$. Hence AA^t is symmetric.

5. Show that any real $n \times n$ matrix can be written as the sum of a symmetric and a skew-symmetric matrix.

Solution: Let $A = (a_{ij})$ be any $n \times n$ matrix.

Let $$b_{ij} = \frac{a_{ij} + a_{ji}}{2} \text{ and } c_{ij} = \frac{a_{ij} - a_{ji}}{2}$$

Then $$a_{ij} = b_{ij} + c_{ij} \text{ and } A = B + C.$$

The matrix $B = (b_{ij})$ is symmetric and $C = (c_{ij})$ is skew symmetric.

6. If A is symmetric and B is skew-symmetric, investigate whether AB is (*i*) symmetric, (*ii*) skew-symmetric.

Solution:

$$A = A^t \quad \text{and} \quad B = -B^t$$

Now $$(AB)^t = B^t A^t = -BA.$$

Thus AB is neither symmetric nor skew-symmetric.

7. Show that all $n \times n$ symmetric matrices over F form a vector space of dimension $\frac{n(n+1)}{2}$ over F. What is the dimension of the space of skew-symmetric matrices?

Solution: Let $A = (a_{ij})$ be an $n \times n$ symmetric matrix. Then $a_{ji} = a_{ij}$, $i \neq j$. Thus the number of independent entries are $a_{ij} (i < j)$ and a_{ii}, $1 \leq i \leq n$ and these are

$\frac{n(n-1)}{2} + n = \frac{n(n+1)}{2}$ in number. Hence the space as dimension $\frac{n(n+1)}{2}$.

For skew-symmetric matrices $a_{ii} = 0$, $1 \le i \le n$ and $a_{ji} = -a_{ij}$. Thus the number of linearly independent entries are $\frac{n(n-1)}{2}$ which is also the dimension of the space of skew-symmetric matrices.

8. Show that it is impossible to find real $n \times n$ matrices A and B such that $AB - BA = I_n$.

Solution: Suppose $AB - BA = I_n$, then

$$Tr(AB - BA) = Tr(I_n),$$

$$T(AB) - Tr(BA) = Tr(I_n)$$

i.e., $0 = n$, a contradiction.

9. Let $a \in F$, $a \neq 0$ and $\theta : M_n(F) \to F$ be the mapping defined by $\theta(A) = a \operatorname{Tr}(A)$. Show that θ is a linear transformation. What is the kernel of θ?

Solution: θ is linear because $\operatorname{Tr}(A)$ is linear.

$$\operatorname{Ker} \theta : \{A \mid \operatorname{Tr}(A) = 0\}.$$

10. Let $\theta : M_n(F) \to F$ be a linear transformation such that $\theta(AB) = \theta(BA)$ for all $A, B \in M_n(F)$. Show that there exists some $a \in F$ such that $\theta(A) = a \operatorname{Tr}(A)$ for all $A \in M_n(F)$.

Solution: Let E_{ij} be the elementary matrix with 1 at the $(i-j)$th place and zero everywhere else.

$$\theta(E_{ij} E_{jk}) = \theta(E_{jk} E_{ij})$$

So $\qquad \theta(E_{ik}) = \theta(E_{jk} E_{ij}) = 0$ where $k \neq i$

Now $\qquad \theta(E_{kk}) = \theta(E_{kj} E_{jk}) = \theta(E_{jk} E_{kj}) = \theta(E_{jj})$

So all $\theta(E_{ii})$ have the same value a.

So $\theta(A) = a \operatorname{Tr}(A)$ is satisfied for all elementary matrices E_{ij}. Since every A is a linear combination of the E_{ij} and both θ and Tr are linear maps, $\theta(A) = a \operatorname{Tr}(A)$ for all A.

11. Let $A \in M_n(F)$ be such that $\operatorname{Tr}(AB) = 0$ for all $B \in M_n(F)$. Show that $A = 0$.

Solution: $\operatorname{Tr}(AE_{11}) = 0$ but AE_{11} is the first column of A and $\operatorname{Tr}(AE_{11}) = a_{11}$. Hence $a_{11} = 0$.

Similarly $\operatorname{Tr}(AE_{ij}) = a_{ij} = 0$ for all i, j. Hence $A = 0$.

12. Let $T : V \to V$ be a linear transformation such that the matrix of T with respect to $\{v_1, v_2, \ldots, v_n\}$ is A. Show that for the linear transformation $S : V^* \to V^*$, where V^* is the dual space of V defined by $S(f) = fT$, $f \in V^*$, the matrix of S with respect to the dual basis $\{v_1^*, v_2^*, \ldots, v_n^*\}$ is A^t.

Solution: Let $T(v_i) = \sum_j a_{ij} v_j$ so that $A = (a_{ij})$.

Now $S(v_i^*) = v_i^* T$

So $\quad S(v_i^*)(v_k) = v_i^* T(v_k) = v_i^* (\sum_j a_{kj} v_j)$

$$= \sum_j a_{kj} v_i^* (v_j)$$

$$= \sum_j a_{kj} \delta_{ij} = a_{ki}$$

Let $S(v_i^*) = \sum_j b_{ij} v_j^*$ so that

$$S(v_i^*)(v_k) = \sum_j b_{ij} v_j^* (v_k) = \sum_j b_{ij} \delta_{jk}$$

$$= b_{ik}.$$

Hence $\quad b_{ik} = a_{ki}$ for all *i, k, i.e.* $B = A^t$.

5.4 RANK

1. For each of the following matrices over reals, find the rank. Which of them are non-singular?

(*i*) $\begin{pmatrix} 1 & 0 \\ 2 & 0 \end{pmatrix}$ $\qquad$ (*ii*) $\begin{pmatrix} 2 & 3 & -1 \\ -1 & 0 & 4 \\ 4 & 5 & 8 \end{pmatrix}$

(*iii*) $\begin{pmatrix} 7 & -1 & 0 & 5 \\ 0 & 1/2 & 1 & 1/3 \\ -3 & -1 & 1/2 & 4 \\ -1 & 0 & 1/2 & 0 \end{pmatrix}$

Solution: (*i*) rank 1. Singular

(*ii*) rank 3. Non-singular

(*iii*) rank 4. Non-singular.

2. Show that for any $A, B \in M_n(F)$, rank $(A + B)$ is not related to rank (A) + rank (B), by giving an example in each of the following cases:

(*i*) rank $(A + B) <$ rank A + rank B.

(*ii*) rank $(A + B) =$ rank A + rank B.

Solution:

(*i*) Let $A = \begin{pmatrix} 1 & 0 \\ 0 & 0 \end{pmatrix}$ and $B = \begin{pmatrix} -1 & 0 \\ 0 & 0 \end{pmatrix}$ $A + B = 0$ and rank $(A + B) = 0$. But rank $(A) =$ rank $(B) = 1$.

(*ii*) $A = \begin{pmatrix} 1 & 0 \\ 0 & 0 \end{pmatrix}$ and $B = \begin{pmatrix} 0 & 0 \\ 0 & 1 \end{pmatrix}$, $A + B = I$ rank $(A + B) = 2$, rank $(A) =$ rank $(B) = 1$.

3. Which matrices have rank zero?

Solution: Only the zero matrix has rank zero.

4. Let $T: V \to V$ be a linear map such that rank $(T) = 1$. Show that $T^2 = aT$ for some $a \in F$. Is the converse true? Why?

Solution: Since rank $(T) = 1$, $T(V)$ is one dimensional, say $T(V) = \{w\}$. Let $T(w) = aw$. Then for any $v \in V$ let $T(v) = bw$. Then $T^2(v) = bT(w) = baw = abw = aT(v)$.

Thus $\qquad T^2 = aT.$

The converse in not true. Consider the 3×3 idempotent matrix

$$E = \begin{pmatrix} 1 & 0 & 0 \\ 0 & 1 & 0 \\ 0 & 0 & 0 \end{pmatrix}.$$

Then $E^2 = E$ but rank (E) is not 1 but 2.

5. Let $T: V \to V$ be a linear map and $n(T) = \dim(\ker T)$ {*nullity of T*}. Show that $r(T) + n(T) = \dim V$.

Solution: $T(V) = V/\ker T$ so that

$$\dim T(V) = \dim V - \dim(\ker T) = \dim V - n(T)$$

i.e., $$r(T) + n(T) = \dim V.$$

6. Let $T, T': V \to V$ be linear maps. Show that $n(TT') \geq \text{Max}\{n(T), n(T')\}$

Solution: $$r(TT') \leq \text{Min}(r(T), r(T'))$$

Since $$n(T) = n - r(T)$$

$$n(TT') \geq \text{Max}\{n(T), n(T')\}.$$

7. Show that for any $A \in Mn(F)$, rank A = rank A^t.

Solution: $$\text{rank } A = \text{row rank } A$$

$$= \text{column rank of } A^t = \text{rank } A^t$$

8. Show that $A = \begin{pmatrix} 1 & -2 & 2 \\ 0 & 1 & 1 \\ 3 & 0 & 1 \end{pmatrix}$ is non-singular and express it as a product of elementary matrices.

Solution: A can be reduced to the identity matrix by the following operations in that order. Column operations $P'_{21}(2)$, $P'_{31}(-2)$ and row operations $P_{31}(-3)$, $P_{32}(-6)$, $P_{23}(11)$ and $P_3(-1/11)$. Hence $PAQ = I$.

where $\qquad P = E_3(-1/11)\, E_{23}(11)\, E_{32}(-6)\, E_{31}(-3)$

and $\qquad Q = E'_{21}(2)\, E'_{31}(-2).$

Hence $A = P^{-1} Q^{-1}$ can be expressed as a product of elementary matrices using the fact $E_{ij}(a)^{-1} = E_{ij}(-a)$, $E_{ij}'(a)^{-1} = E_{ij}'(-a)$ and $E_i(a)^{-1} = E_i(a^{-1})$.

9. Reduce $A = \begin{pmatrix} 4 & 0 & 1 \\ 2 & 3 & 6 \\ 6 & -3 & -1 \end{pmatrix}$ to the unit matrix by elementary operations and find A^{-1}.

Solution: The following row and column operations done in that order bring A to I.

$$P_{12}, P_{21}(-2), P_{31}(-3), P_{12}(1/2), P_{32}(-2)$$
$$P'_{31}(-1/4), P'_{32}(11/6), P_1(1/2), P_2(-1/6) \text{ and } P_3(1/3).$$

To get the inverse of A, perform the same operations in the same order on the identity matrix. This gives

$$A^{-1} = \frac{1}{36}\begin{pmatrix} 15 & 38 & -24 \\ -3 & -10 & 12 \\ -3 & -22 & 12 \end{pmatrix}^{+}$$

10. Test for equivalence, the matrices

$$A = \begin{pmatrix} 0 & 0 & 0 \\ 0 & 0 & 0 \\ 0 & 0 & 2 \end{pmatrix} \text{ and } B = \begin{pmatrix} 1 & 1 & 3 \\ -1/2 & 1 & 0 \\ 0 & 1/2 & 4 \end{pmatrix}$$

Solution: Rank $A = 1$ and rank $B = 3$ and hence they are not equivalent.

11. Find the canonical form under equivalence for the matrix

$$A = \begin{pmatrix} 3 & 1 & -2 \\ 2 & 0 & -5 \\ 4 & 2 & 1 \end{pmatrix}$$

Solution: Rank $(A) = 2$. Hence the canonical form of A is $\begin{pmatrix} 1 & 0 & 0 \\ 0 & 1 & 0 \\ 0 & 0 & 0 \end{pmatrix}$.

12. If A and B are equivalent matrices investigate whether the following are equivalent?

(*i*) A^t and B^t.

(*ii*) A^2 and B^2.

(*iii*) AB and BA.

Solution:

(*i*) Yes. If $B = PAQ$, $B^t = Q^t A^t P^t$

(*ii*) No. Example.

Let $A = \begin{pmatrix} 0 & 1 \\ 0 & 0 \end{pmatrix}$ and $B = \begin{pmatrix} 1 & 0 \\ 0 & 0 \end{pmatrix}$. Then B is equivalent to A as it is obtained by interchanging columns.

Now $A^2 = \begin{pmatrix} 0 & 0 \\ 0 & 0 \end{pmatrix}$ and $B^2 = B \neq 0$. Hence A^2 and B^2 are not equivalent.

(*iii*) No. Example.

Let A and B be as in the above example. Then A and B are equivalent. But $AB = \begin{pmatrix} 0 & 0 \\ 0 & 0 \end{pmatrix}$ and $BA = \begin{pmatrix} 0 & 1 \\ 0 & 0 \end{pmatrix}$. Hence AB and BA are not equivalent.

13. Show that for any A, A and A^t are equivalent.

Solution: If rank $A = r$, rank $A^t = r$. Hence both have canonical form

$$\begin{pmatrix} 1 & 0 & \dots & 0 \\ 0 & 1 & \dots & 0 \\ \dots & \dots & \dots & \dots \\ \dots & 0 & 1 & \dots \\ \dots & \dots & \dots & \dots \\ 0 & 0 & \dots & 0 \end{pmatrix}$$ with 1 occurring r times on the diagonal. Hence they are equivalent.

14. Let F_{ij} be the matrix with 1 at the $(i–j)$th place and zero everywhere else. Show that all the E_{ij} are equivalent to each other.

Solution: Since all the E_{ij}'s are all of rank n, they are equivalent to each other, having the same canonical form.

5.5 LINEAR EQUATIONS

1. Find the dimensions of the space of solutions of

$$2x_1 + x_2 - 3x_3 - x_4 = 0$$
$$x_1 + 2x_3 = 0$$
$$3x_1 - x_2 - x_3 + x_4 = 0.$$

Solution: The coefficient matrix in this case is

$$A = \begin{pmatrix} 2 & 1 & -3 & -1 \\ 1 & 0 & 2 & 0 \\ 3 & -1 & -1 & 1 \end{pmatrix}, \text{ rank } (A) = 3.$$

Hence the required dimension is $4 - 3 = 1$.

2. Find a basis for the space of solutions of

$$x_1 - x_2 + 2x_3 = 0$$
$$3x_1 + 4x_2 + 5x_3 = 0$$
$$x_1 + x_2 + x_3 = 0.$$

Solution: The coefficient matrix in this case is

$$A = \begin{pmatrix} 1 & -1 & 2 \\ 3 & 4 & 5 \\ 1 & 1 & 1 \end{pmatrix}, \text{ rank } (A) = 3.$$

Hence the solution space is $\{0\}$.

3. Solve the system of non-homogeneous linear equations. Does it have a unique solution? Why?

$$x_1 - 2x_2 + x_3 = 1$$
$$2x_1 - 5x_2 + 2x_3 = 2$$
$$x_1 + x_2 - x_3 = -1.$$

Solution: The coefficient matrix in this case is $A = \begin{pmatrix} 1 & -2 & 1 \\ 2 & -5 & 2 \\ 1 & 1 & -1 \end{pmatrix}$, rank (A) $= 3$. It is also the rank of augmented matrix $A' = \begin{pmatrix} 1 & -2 & 1 & 1 \\ 2 & -5 & 2 & 2 \\ 1 & 1 & -1 & -1 \end{pmatrix}$. Hence the system has a solution. The solution is unique and is given by

$$\begin{pmatrix} x_1 \\ x_2 \\ x_3 \end{pmatrix} = -A^{-1} \begin{pmatrix} 1 \\ 2 \\ -1 \end{pmatrix}.$$

4. Let $A = (a_{ij})$ be an $n \times n$ matrix over reals. If the column-vectors are linearly independent over reals, show that they are also linearly independent over complex numbers.

Solution: Let $C_1, C_2, \ldots, C_n$ be the columns of A.

Let $\sum_k \lambda_k C_k = 0$ where $\lambda_k = a_k + ib_k$.

Then $\sum_k (a_k + ib_k) C_k = 0$, *i.e.* $\sum_k a_k C_k + i \sum_k b_k C_k = 0$.

Since C_k's are real, $\sum_k a_k C_k = 0 = \Sigma\, b_k C_k$. Since C_k are linearly independent over reals, $a_k = 0 = b_k$ for all k. Hence $\lambda_k = 0$ for all k.

5. Show that a system of m non-homogeneous linear equations in n unknowns $(m < n)$ has either no solution or has many solutions.

Solution: If the rank of the coefficient matrix is not equal to the rank of the augmented matrix there is no solution. Otherwise the set of solutions is obtained by adding to one-solution, $m - n$ linearly independent solutions. Since $m - n > 0$, there are infinitely many solutions.

6. Show that $ax_1 + bx_2 = 0$ and $cx_1 + dx_2 = 0$ has only a trivial solution if, and only if, $ad - bc \neq 0$.

Solution: $\begin{pmatrix} a & b \\ c & d \end{pmatrix} \begin{pmatrix} x_1 \\ x_2 \end{pmatrix} = \begin{pmatrix} 0 \\ 0 \end{pmatrix}$ has only trivial solution if, and only if, $\begin{pmatrix} a & b \\ c & d \end{pmatrix}$ is non-singular, *i.e.* $ad - bc \neq 0$.

7. Determine values of a for which the following system of equations:

(*i*) has no solution,

(*ii*) has a unique solution,

(*iii*) has infinitely many solutions:

$$x_1 + x_2 + x_3 = 2$$
$$x_1 + 2x_2 + x_3 = -2$$
$$x_1 + x_2 + (a-5)x_3 = a.$$

Solution: (*i*) The equations may be written as

$$\begin{pmatrix} 1 & 1 & 1 \\ 1 & 2 & 1 \\ 1 & 1 & a-5 \end{pmatrix} \begin{pmatrix} x_1 \\ x_2 \\ x_3 \end{pmatrix} = \begin{pmatrix} 2 \\ -2 \\ a \end{pmatrix}$$

(*ii*) For unique solution the coefficient matrix

$$A = \begin{pmatrix} 1 & 1 & 1 \\ 1 & 2 & 1 \\ 1 & 1 & a-5 \end{pmatrix} \text{ should be non-singular.}$$

Now let $(A) = a - 6$ and if $a \neq 6$, the solution is unique.

(*i*) Let $a = 6$. $A = \begin{pmatrix} 1 & 1 & 1 \\ 1 & 2 & 1 \\ 1 & 1 & 1 \end{pmatrix}$ has rank 2. The augmented matrix is

$$\begin{pmatrix} 1 & 1 & 1 & 2 \\ 1 & 2 & 1 & -2 \\ 1 & 1 & 1 & 6 \end{pmatrix} \text{ has rank 3. Hence there is no solution.}$$

(*iii*) The case of infinitely solutions does not arise for any value of a.

5.6 DETERMINANTS

1. Show that if $A \in M_n(F)$, then $D(aA) = a^n D(A)$.

Solution: Let $A = (C_1, \ldots, C_n)$ be the columns of A. Then $aA = (aC_1, \ldots, aC_n)$. Using linearity on the columns $D(aA) = a^n D(A)$.

2. Show that a nilpotent matrix has zero determinant and a non-singular idempotent matrix is I_n.

Solution: (*i*) Let A be nilpotent so that $A^m = 0$.

$$0 = \det(A^m) = (\det(A))^m. \text{ So } \det(A) = 0.$$

(*ii*) Let A be non-singular and $A^2 = A$. Multiplying by A^{-1}, $A = I$.

3. Show that a skew-symmetric matrix of odd order is singular.

Solution: Let $A^t = -A$

$$\det(A^t) = \det(-A) = (-1)^n \det(A).$$

So $\quad \det(A) = (-1)^n \det(A)$.

If n is odd, $\det(A) = 0$. Hence A is singular.

4. Show that the determinant of a triangular matrix is the product of its diagonal elements. Deduce that a triangular matrix is non-singular if, and only if, each diagonal element is non-zero.

Solution: Induction on n. $n = 2$, $A = \begin{pmatrix} \lambda_1 & 0 \\ * & \lambda_2 \end{pmatrix}$ has determinant $\lambda_1\lambda_2$.

Let $A = \begin{pmatrix} \lambda_1 & 0 & \dots & 0 \\ & \lambda_2 & & \\ & & \ddots & \\ * & & \dots & \lambda_n \end{pmatrix}$. Then by expanding along the elements of the first row, $\det(A) = \lambda_1 \det(A_1)$, where $A_1 = \begin{pmatrix} \lambda_2 & & 0 \\ & \ddots & \\ * & & \lambda_n \end{pmatrix}$. By induction $\det(A_1) = \lambda_2 \dots \lambda_n$ so that $\det(A) = \lambda_1\lambda_2 \dots \lambda_n$.

In particular A is non-singular if and only if $\det(A) \neq 0$, *i.e.* $\lambda_i \neq 0$ for each i.

5. Calculate the determinant of the matrix

$$\begin{pmatrix} 1 & 1 & \dots & 1 \\ x_1 & x_2 & \dots & x_n \\ \dots\dots & \dots\dots & \dots\dots & \dots\dots \\ x_1^{n-1} & x_2^{n-1} & \dots & x_n^{n-1} \end{pmatrix}$$

Solution: Consider the polynomial

$$V_{n-1}(x) = \begin{vmatrix} 1 & 1 & \dots & 1 & 1 \\ x_1 & x_2 & \dots & x_{n-1} & x \\ \dots & \dots & \dots & \dots & \dots \\ x_1^{n-1} & x_2^{n-1} & \dots & x_{n-1}^{n-1} & x^{n-1} \end{vmatrix}$$

This is a polynomial in x of degree $n - 1$ with $x_1, x_2 \dots n_{n-1}$ as roots and leading coefficient $V_{n-2}(x_{n-1})$.

Hence $\quad V_{n-1}(x) = V_{n-2}(x_{n-1})(x - x_1)(x - x_2) \dots (x - x_{n-1})$.

Putting $x = x_n$, we get

$$V_{n-1}(x_n) = V_{n-2}(x_{n-1})(x_n - x_1)(x_n - x_2) \dots (x_n - x_{n-1}).$$

By repetition, we get $V_{n-1}(x_n) = \prod\limits_{\substack{i,j \\ n \geq i > j \geq 1}} (x_i - x_j)$.

6. Show that $A \in M_n(F)$ has rank r if, and only if, (*i*) every $(r + 1) \times (r + 1)$ submatrix of A has determinant zero, and (*ii*) there exists and $r \times r$ submatrix of A with non-zero determinant.

Solution: Let rank $A = r$. Then for any $(r + 1) \times (r + 1)$ submatrix of A, the corresponding rows of A are linearly dependent as rank $A = r$. Hence the rows of the submatrix are also linearly dependent, *i.e.,* its determinant is zero.

Since rank $A = r$, there exists r linearly independent rows. The corresponding submatrix with these as rows has rank r so that it has r linearly independent columns. The matrix of columns from the submatrix which is an $r \times r$-matrix with determinant non-zero. Conversely assume conditions (*i*) and (*ii*). If rank $A = \rho$, then ρ satisfies (*i*) and (*ii*) and hence $\rho = r$.

7. Let $A, B \in M_n(F)$ with $D(B) \neq 0$. Show that $D(A + tB)$ is a polynomial in t with the highest coefficient equal to $D(B)$ and with constant term $D(A)$.

Solution: By the expansion rule for the determinant

$$D(A + tB) = D(a_{ij} + tb_{ij})$$

is a polynomial in t of deg n say $P_0t^n + P_1t^{n-1} + \ldots + P_n$.

By putting $t = 0$ in this polynomial, we get $P_n = \det(A)$.

Let $\tau = \dfrac{1}{t}$. Then $P_0 + P_1\tau + \ldots + P_n\tau^n = D(\tau a_{ij} + b_{ij})$

Put $\tau = 0$, we get $P_0 = D(B)$.

8. Let M_1 and M_2 be $r \times r$ and $s \times s$ matrices. Show that if $M = \begin{pmatrix} M_1 & 0 \\ 0 & M_2 \end{pmatrix}$ is the $(r + s) \times (r + s)$ matrix, then

$$D(M) = D(M_1)\, D(M_2).$$

Solution: By the expansion rule for the determinant of M, any non-zero term in the expansion must be a product of a non-zero term in the expansion of determinant of M_1, and a non-zero term in the expansion of determinant of M_2 and vice versa. With the signs of the terms also matching in each case, we have $D(M) = D(M_1)\, D(M_2)$.

9. Solve the following equations by Cramer's rule

$$x_1 + x_2 - 2x_3 = 1$$
$$2x_1 - 7x_3 = 3$$
$$x_1 + x_2 - x_3 = 5$$

Solution: By Cramer's rule

$$\frac{x_1}{\begin{vmatrix} 1 & 1 & -2 \\ 3 & 0 & -7 \\ 5 & 1 & -1 \end{vmatrix}} = \frac{x_2}{\begin{vmatrix} 1 & 1 & -2 \\ 2 & 3 & -7 \\ 1 & 5 & -1 \end{vmatrix}}$$

$$= \frac{x_3}{\begin{vmatrix} 1 & 1 & 1 \\ 2 & 0 & 3 \\ 1 & 1 & 5 \end{vmatrix}} = \frac{1}{\begin{vmatrix} 1 & 1 & -2 \\ 2 & 0 & -7 \\ 1 & 1 & -1 \end{vmatrix}}$$

This gives the values of x_1, x_2 and x_3.

10. Compute $D(A)$, where $A = \begin{pmatrix} 0 & I_n \\ -I_n & 0 \end{pmatrix}$, where I_n is the unit $n \times n$ matrix.

Solution: By interchanging, the first and $(n + 1)$th column, 2nd and $(n + 2)$th column, etc. we get the matrix $B = \begin{pmatrix} I_n & 0 \\ 0 & -I_n \end{pmatrix}$.

Then $\quad D(A) = (-1)^n D(B).$

But $\quad D(B) = D(I_n) D(-I_n) = 1 \cdot (-1)^n = (-1)^n.$

Then $\quad D(A) = (-1)^{2n} = 1.$

11. Show that the following matrix is non-singular

$$\begin{pmatrix} 1 & 1 & 1 & \dots & 1 \\ 1 & 2 & 2^2 & \dots & 2^{n-1} \\ \dots & \dots & \dots & \dots & \dots \\ 1 & n & n^2 & \dots & n^{n-1} \end{pmatrix}.$$

Solution: This is the transpose of the matrix in Problem 5, where $x_1 = 1$, $x_2 = 2, \dots, x_n = n$. Since the determinant is $\prod\limits_{i<j} (x_i - x_j)$ and since all the x_i's are distinct the determinant is non-zero. Hence it is non-singular.

5.7 MINIMUM POLYNOMIAL

1. Find the minimum polynomials of the following matrices. Find also their in inverses if they are non-singular.

(i) $\begin{pmatrix} 0 & 0 & 0 \\ 1 & 0 & -9 \\ 0 & 1 & 6 \end{pmatrix}$ $\qquad$ (ii) $\begin{pmatrix} 2 & 1 & -1 \\ 3 & 4 & 2 \\ 0 & 0 & 1 \end{pmatrix}$.

Solution: (i) $A^2 = \begin{pmatrix} 0 & 0 & 0 \\ 0 & -9 & -54 \\ 1 & 6 & 27 \end{pmatrix}$ $\quad 6A = \begin{pmatrix} 0 & 0 & 0 \\ 6 & 0 & -54 \\ 0 & 6 & 36 \end{pmatrix}$

So $\quad A^2 - 6A + 9 = \begin{pmatrix} 9 & 0 & 0 \\ -6 & 0 & 0 \\ 1 & 0 & 0 \end{pmatrix}$

$A(A^2 - 6A + 9)$ is the zero matrix.

Hence the minimum polynomial is $x^3 - 6x^2 + 9x$.

(ii) $(A - I)^2 = \begin{pmatrix} 4 & 4 & 1 \\ 12 & 12 & 3 \\ 0 & 0 & 0 \end{pmatrix}$ and $A - 5I = \begin{pmatrix} -3 & 1 & -1 \\ 3 & -1 & 2 \\ 0 & 0 & -4 \end{pmatrix}$

$(A - I)^2 (A - 5I)$ is the zero matrix.

Hence the minimum polynomial is

$$(x-1)^2(x-5) = x^3 - 7x^2 + 11x - 5.$$

Case (*i*) is singular matrix.

Case (*ii*) is singular. $A^{-1} = \frac{1}{5}(A^2 - 7A + 11I)$.

2. Let $V = F_3[x]$ be the space of all polynomials of degree utmost 3 and let $T : V \to V$ be the linear transformation given by $T(f) = f'$. Find the minimum polynomial of T.

Solution: The basis for V is $\{1, x, x^2, x^3\}$ and $T(1) = 0$, $T(x) = 1$, $T(x^2) = 2x$ and $T(x^3) = 3x^2$. Hence the matrix of T is

$$A = \begin{pmatrix} 0 & 0 & 0 & 0 \\ 1 & 0 & 0 & 0 \\ 0 & 2 & 0 & 0 \\ 0 & 0 & 3 & 0 \end{pmatrix}$$

By direct computation $A^4 = 0$ and $A^3 \neq 0$. Hence the minimum polynomial of T is x^4.

3. Let V be a four dimensional vector space with basis v_1, v_2, v_3 and v_4. Let $T : V \to V$ be the linear map defined by $T(v_1) = v_2$, $T(v_2) = v_3$, $T(v_3) = v_4$ and $T(v_4) = -5v_1 - 2v_2 - \frac{1}{2}v_3 - v_4$. Find the minimum polynomial of T.

Solution: The matrix of T in this case is

$$A = \begin{pmatrix} 0 & 1 & 0 & 0 \\ 0 & 0 & 1 & 0 \\ 0 & 0 & 0 & 1 \\ -5 & -2 & -\frac{1}{2} & -1 \end{pmatrix}$$

This is the companion matrix of the polynomial

$$q(x) = x^4 - x^3 - \frac{1}{2}x^2 - 2x - 5.$$

Hence its characteristic polynomial is $q(x)$ (Problem 4, 5.9). This is also the minimum polynomial of A, as $q(x)$ has no repeated roots by showing that $q(x)$ and $q'(x)$ has g.c.d. unit. Thus the minimum and characteristic polynomials of A coincide (Problem 13, 5.8). The minimum polynomial is equal to $x^4 - x^3 - \frac{1}{2}x^2 + 2x - 5$.

4. What is the minimum polynomial of a diagonal matrix with diagonal elements $d_1, d_2, \ldots, d_n$?

Solution: $A - d_1I$ has the first diagonal entry zero, $A - d_2I$ has the second diagonal entry zero and so on. Hence the product $(A - d_1I) \ldots (A - d_nI)$ is the zero matrix. Hence the minimum polynomial is

$$(x - d_1)(x - d_2) \ldots (x - d_n).$$

5. What is the minimum polynomial of a non-zero

(*i*) nilpotent matrix

(*ii*) idempotent matrix?

Solution: (*i*) If $A^m = 0$ and $A^{m-1} \neq 0$, the minimum polynomial is x^m.

(*ii*) $A^2 = A$ so that A satisfies $x(x-1)$. If $A = I$, the minimum polynomial is $x - 1$. Otherwise it is $x(x-1)$.

6. Show that for any $A \in M_n(F)$ both A and A^t have the same minimum polynomial.

Solution: Let A have the minimum polynomial

$$m(x) = a_0 + a_1 x + \ldots + a_m x^m.$$

Then $$I + a_1 A + \ldots + a_m A^m = 0.$$

By taking transpose, $I + a_1 A^t + \ldots + a_m (A^t)^m = 0$. Hence A^t also satisfies $m(x)$. It is also the minimum polynomial as any polynomial satisfied by A^t is also satisfied by A.

7. Let $A \in M_n(F)$ with $A = \begin{pmatrix} B & 0 \\ 0 & C \end{pmatrix}$, where $B \in M_r(F)$ and $C \in M_s(F)$ such that $r + s = n$ and 0 is the zero $s \times r$ and $r \times s$ matrices. Show that the minimum polynomial of A is the l.c.m. of the minimum polynomials of B and C.

Solution: Let $f(x)$ be the minimum polynomial of B and $g(x)$ the minimum polynomial of C. Let $h(x)$ be the l.c.m. of $f(x)$ and $g(x)$. Then $h(B) = 0 = h(C)$.

Now $h(A) = \begin{pmatrix} h(B) & 0 \\ 0 & h(C) \end{pmatrix} = 0$. So the minimum polynomial $\theta(x)$ of A divides $h(x)$. Now we claim that $h(x)$ divides $\theta(x)$. $\theta(A) = \begin{pmatrix} \theta(B) & 0 \\ 0 & \theta(C) \end{pmatrix} = 0$. Hence $\theta(B) = 0 = \theta(C)$. Here the minimum polynomials of B and C divide $\theta(x)$ *i.e.*, $f(x)$ and $g(x)$ divide $\theta(x)$. Hence their l.c.m. $h(x)$ divides $\theta(x)$. Thus $h(x)\ \theta(x)$.

8. Let A be a real symmetric matrix and $f(x)$ a polynomial with real coefficients. Show that $f(A)$ is also real symmetric.

Solution: Since f has real coefficients

$f(A^t) = f(A)^t$. But $A = A^t$ so that

$f(A) = f(A)^t$. Hence $f(A)$ is symmetric.

9. Let $T : V \to V$ be a linear transformation. Assume that there exists an element $v_0 \in V$ such that, any $v \in V$ can be written as $v = f(T)\, v_0, f(x) \in F[x]$. Let $q(x) = a_0 + a_1 x + \ldots + x^m$ be the minimum polynomial of T. Show that the matrix of T with respect to suitable basis of V can be put in the form

$$C(q(x)) = \begin{pmatrix} 0 & 1 & 0 & \ldots & 0 \\ 0 & 0 & 1 & \ldots & 0 \\ \ldots & \ldots & \ldots & \ldots & \ldots \\ 0 & 0 & \ldots & & 1 \\ -a_0 & -a_1 & \ldots & \ldots & -a_{m-1} \end{pmatrix}$$

Solution: Choose as basis of V, v_0, Tv_0, ... $T^{m-1}\ v_0$. These are linearly independent, for if

$$a_0 v_0 + a_1\ Tv_0 + \ldots + a_{m-1}\ T^{m-1}\ v_0 = 0$$

$$(a_0 + a_1 T + \ldots + a_{m-1}\ T^{m-1})\ v_0 = 0$$

$(a_0 + \ldots + a_{m-1}\ T^{m-1})\ v = 0$ for any v as $v = f(T)\ v_0$, *i.e.* T satisfies a polynomial of deg $< m$, a contradiction. Also these elements generate V, as any $v \in V$ can be written as $v = f(T)\ v_0$.

Write $\quad f(x) = \theta(x)\ q(x) + r(x)$, deg $r(x) < m$

Then $\quad v = f(T)\ v_0 = r(T) v_0$ is a linear combination of v_0, Tv_0, ..., $T^{m-1}\ v_0$.

With respect to this basis the matrix of T has the required form.

10. Give example of two $n \times n$ matrices which have the same minimum polynomial, but are not similar to each other.

Solution: $A = \begin{pmatrix} 1 & 1 \\ 0 & 0 \end{pmatrix}$ and $B = \begin{pmatrix} 0 & 1 \\ 1 & 1 \end{pmatrix}$ have the same minimum polynomial $x(x-1)$ but A is not similar to B as rank $A = 1$ and rank $B = 2$.

5.8 CHARACTERISTIC ROOTS

1. Show that the characteristic roots of

$$A = \begin{pmatrix} 0 & 1 & 0 & 0 \\ 0 & 0 & 1 & 0 \\ 0 & 0 & 0 & 1 \\ 1 & 0 & 0 & 0 \end{pmatrix}$$

are the fourth roots of unity.

Solution: $$D(\lambda I - A) = \begin{vmatrix} \lambda & -1 & 0 & 0 \\ 0 & \lambda & -1 & 0 \\ 0 & 0 & \lambda & -1 \\ -1 & 0 & 0 & \lambda \end{vmatrix} = \lambda^4 - 1.$$

Hence the roots are the fourth root of unity.

2. Show that zero is a characteristic root of A if, and only if, A is singular.

Solution: If A is singular, $AX = 0 = 0X$ for some non-zero X. Hence 0 is a characteristic root. Conversely assume that zero is a characteristic root of A, so that the characteristic polynomial has constant term zero. But the constant term is the determinant of A (upto sign). So $D(A) = 0$, *i.e.*, A is singular.

3. If A is non-singular, show that the characteristic roots of A^{-1} are the inverses of the characteristic roots of A.

Solution: Let λ be a characteristic root of A. Then $\lambda \neq 0$ and $AX = \lambda X$, $X \neq 0$. Now $A^{-1}(AX) = A^{-1}(\lambda X)$, *i.e.*, $X = \lambda A^{-1}(X)$. So $A^{-1}(X) = \lambda^{-1} X$. Hence λ^{-1} is a characteristic root of A^{-1}. Converse is obtained by reversing the steps.

4. Show that A is nilpotent if, and only if, all its characteristic roots are zero.

Solution: If A is nilpotent, its minimum polynomial has the form x^m and all the root are zero. Hence all the characteristic roots are zero. Conversely, if all the characteristic roots are zero, the characteristic polynomial is x^n and by Cayley-Hamilton theorem $A^n = 0$. Hence A is nilpotent.

5. Show that any nilpotent matrix has trace zero.

Solution: Since A is nilpotent, $A^m = 0$ for some m. Hence all the characteristic roots are zero and the characteristic polynomial is $P(x) = x^n$. Since trace is the coefficient (upto sign) of x^{n-1}, Tr. $(A) = 0$.

6. Show that if A is nilpotent, then A is similar to a triangular matrix, all of whose diagonal entries are zero.

Solution: All the characteristic roots are zero and lie in the base field F. Hence A is similar to a triangular matrix whose diagonal entries are the characteristic roots of A, *i.e.*, all the diagonal entries are zero.

7. Let $A \in M_n(F)$ be such that the sum of the terms of A along each row is 1. Show that 1 is a characteristic root of A.

Solution: $A \begin{pmatrix} 1 \\ 1 \\ \cdots \\ 1 \end{pmatrix} = \begin{pmatrix} 1 \\ 1 \\ \cdots \\ 1 \end{pmatrix}$ since the sum along each row is 1.

Hence 1 is a characteristic root with characteristic vector $\begin{pmatrix} 1 \\ 1 \\ \cdots \\ 1 \end{pmatrix}$.

8. Show that any matrix similar to a diagonal matrix is similar to its transpose.

Solution: Let A be similar to $\begin{pmatrix} d_1 & & \\ & \ddots & \\ & & d_n \end{pmatrix}$. Then A^t is similar to

$$\begin{pmatrix} d_1 & & \\ & \ddots & \\ & & d_n \end{pmatrix}^t = \begin{pmatrix} d_1 & & \\ & \ddots & \\ & & d_n \end{pmatrix} = A.$$

9. Show that for any $A, B \in M_n(F)$, AB and BA have the same characteristic polynomial. In particular, they have the same characteristic roots.

Solution: Case (i): A is non-singular. The $BA = A^{-1} (AB) A$. Hence BA is similar to AB and hence they have the same characteristic polynomial. Similarly, the case when B is non-singular.

Case (ii): A is singular and B is singular. In this case both AB and BA are singular. Hence both have zero as characteristic root. We show that the non-zero characteristic roots coincide. Let $\lambda \neq 0$ be a characteristic root of AB so that $ABv = \lambda v$, $v \neq 0$. Let $Bv = w$ so that $Aw \neq 0$. $BAw = BABv = B(\lambda v) = \lambda Bv = \lambda w$, $\lambda \neq 0$. Hence λ is a characteristic root of BA. The converse can be argued similarly.

10. Show that AB and BA need not have the same minimum polynomial.

Solution: Let

$$A = \begin{pmatrix} 0 & 1 \\ 0 & 0 \end{pmatrix} \quad \text{and} \quad B = \begin{pmatrix} 1 & 0 \\ 0 & 0 \end{pmatrix}$$

Then
$$AB = \begin{pmatrix} 0 & 0 \\ 0 & 0 \end{pmatrix} \quad \text{and} \quad BA = \begin{pmatrix} 0 & 1 \\ 0 & 0 \end{pmatrix}.$$

Minimum polynomial of AB is x and minimum polynomial of BA is x^2.

11. Let $A, B \in M_n(F)$ be such that $AB = BA$. Show that A and B have a common characteristic vector.

Solution: Let λ be a characteristic root of A and W the set of corresponding characteristic vectors so that $Av = \lambda v$, for all $v \in W$. Since B commutes with A, $B(W) \subset W$, for if $v \in W$, then $A(Bv) = BA(v) = B\lambda v = \lambda Bv$, so that $Bv \in W$. Then $B \mid W : W \to W$ is a linear map. Take any characteristic vector of this linear map say $w_0 \in W$, so that $Bw_0 = \mu w_0$. Then w_0 is a common characteristic vector of both A and B corresponding to characteristic roots λ and μ respectively.

12. Let λ be a characteristic root of A. Show that the set of all characteristic vectors corresponding to λ form a subspace.

Solution: If
$$AX = \lambda X \quad \text{and} \quad AY = \lambda Y,$$
then
$$A(X - Y) = \lambda(X - Y) \quad \text{and} \quad A(aX) = \lambda(aX).$$

Thus all characteristic vectors corresponding to λ form a subspace.

13. Let $A \in M_n(F)$. Show that the characteristic polynomial of A divides a powers of the minimum polynomial of A.

Solution: The characteristic polynomial and the minimum polynomial have the same roots in some extension field of F.

Let
$$P_A(t) = (t - \lambda_1)^{\alpha_1} \ldots (t - \lambda_k)^{\alpha_k}$$

and
$$m(t) = (t - \lambda_1)^{\beta_1} \ldots (t - \lambda_k)^{\beta_k}$$

withe $\beta_i \leq \alpha_i$ for each i. Choose large N such that $N\beta_i \geq \alpha_i$ for each i. Then $P_A(t) \mid m(t)^N$ in the extension field. Since both polynomials are in $F[x]$, $P_A(t)$ divides $m(t)^N$ in $F[x]$.

5.9 CANONICAL FORMS

1. Prove that the matrix $A = \begin{pmatrix} 1 & 1 & 1 \\ -1 & -1 & -1 \\ 1 & 1 & 0 \end{pmatrix}$ is nilpotent. Reduce it to triangular form.

Solution: $A^2 = \begin{pmatrix} 1 & 1 & 0 \\ -1 & -1 & 0 \\ 0 & 0 & 0 \end{pmatrix}$ and $A^3 = 0$. Hence A is nilpotent, with minimum polynomial of A, $q(t) = t^3$. Hence the triangular form of A is the Jordan Canonical form of A which is $\begin{pmatrix} 0 & 1 & 0 \\ 0 & 0 & 1 \\ 0 & 0 & 0 \end{pmatrix}$.

2. Reduce to Jordan form, the matrix $A = \begin{pmatrix} 0 & 1 & 0 & 0 \\ 0 & 0 & 2 & 0 \\ 0 & 0 & 0 & 3 \\ 0 & 0 & 0 & 0 \end{pmatrix}$.

Solution: The characteristic polynomial of A is $D(\lambda I - A) = \lambda^4$. Hence $A^4 = 0$, *i.e.*, A is nilpotent. Moreover, by computation $A^3 \neq 0$ so that the minimum polynomial of A is $q(t) = t^4$. Hence the Jordan canonical from of A is given by

$$\begin{pmatrix} 0 & 1 & 0 & 0 \\ 0 & 0 & 1 & 0 \\ 0 & 0 & 0 & 1 \\ 0 & 0 & 0 & 0 \end{pmatrix}.$$

3. Find the rational canonical form of $A = \begin{pmatrix} 1 & 1 & 1 & 1 \\ 0 & 0 & 0 & 0 \\ 0 & 0 & -1 & 0 \\ 0 & -1 & 1 & 0 \end{pmatrix}$.

Solution: The characteristic polynomial of A is $(x-1)\,x\,(x^2+x+1)$ and since the factors are non-square it is also the minimum polynomial of A. Hence the rational canonical form is $C(x) \oplus C(x-1) \oplus C(x^2+x+1)$ where $C(q(x))$ is the companion matrix of $q(x)$. In block matrix notation this can be expressed as

$$\left(\begin{array}{c|c|cc} 0 & & & \\ \hline & 1 & & \\ \hline & & 0 & 1 \\ & & -1 & -1 \end{array}\right)$$

4. Show that the characteristic polynomial of the companion matrix $C(q(x))$ is $q(x)$.

Solution:

$$C(q(x)) = \begin{pmatrix} 0 & 1 & 0 & \dots & 0 \\ 0 & 0 & 1 & \dots & 0 \\ \dots & \dots & \dots & \dots & \dots \\ 0 & 0 & 0 & \dots & 1 \\ -a_0 & -a_1 & \dots & \dots & a_{n-1} \end{pmatrix}$$

where $\qquad q(x) = a_0 + a_1 x + \ldots + a_{n-1} x^{n-1} + x^n.$

Solution: $$D(\lambda I - q(x)) = \begin{vmatrix} \lambda & -1 & 0 & \ldots & & 0 \\ 0 & \lambda & -1 & \ldots & & 0 \\ \ldots & \ldots & \ldots & \ldots & & \ldots \\ 0 & \ldots & \ldots & \lambda & & -1 \\ a_0 & a_1 & \ldots & \ldots & & \lambda + a_{n-1} \end{vmatrix}$$

Add to the first column, λ times the second column, λ^2 times the third column etc. and λ^{n-1} times the last column. Then the determinant becomes

$$\begin{vmatrix} 0 & -1 & 0 & \ldots & & 0 \\ 0 & \lambda & -1 & \ldots & & 0 \\ \ldots & \ldots & \ldots & \ldots & & \ldots \\ 0 & \ldots & \ldots & \ldots & \lambda & -1 \\ q(\lambda) & a_1 & \ldots & \ldots & \ldots & \lambda + a_{n-1} \end{vmatrix}$$

and this is equal to

$$(-1)^{n-1} q(\lambda) \begin{vmatrix} -1 & 0 & \ldots & 0 \\ \lambda & -1 & \ldots & 0 \\ \ldots & \ldots & \ldots & \ldots \\ 0 & \ldots & \ldots & \lambda - 1 \end{vmatrix}$$

$$= (-1)^{n-1} q(\lambda) (-1)^{n-1} = q(\lambda).$$

5. Deduce from the previous problem that the characteristic polynomial of T is the product of all elementary divisors of T.

Solution: Under the Rational Canonical form, the matrix of T is $\begin{pmatrix} R_1 & & 0 \\ & \ddots & \\ 0 & & R_k \end{pmatrix}$,

where each $R_i = \begin{pmatrix} C(q_i(x)^{e_{i1}}) & & \\ & \ddots & \\ & & C(q_i(x)^{e_{in_i}}) \end{pmatrix}$. Then the characteristic polynomial of T is the product of characteristic polynomials of R_i, for each R_i, the characteristic polynomial is the product of the characteristic polynomials of $C(q_i(x)^{e_{ij}})$ which is nothing but $q_i(x)^{e_{ij}}$. Hence the characteristic polynomial of T is the product of all elementary divisors.

6. Let $T : V \rightarrow V$ be a linear map such that all the characteristic roots of T lie in F. Show that T is diagonalisable if, and only if, $(T - \lambda)^n v = 0, v \in V, \lambda \in F$ implies $(T - \lambda)(v) = 0$.

Solution: Since all the characteristic roots lie in F, the minimum polynomial of T is of the form $q(t) = \prod_{i=1}^{s} (t - \lambda_i)^{n_i}$. Assume that $(T - \lambda)^n v = 0$ implies $(T - \lambda) v = 0$. We claim that each $n_i = 1$.

Suppose some n_i say $n_1 \geq 2$. Then $(T - \lambda_1)^{n_1} \ (\prod_{i \neq 1} T - \lambda_i)^{n_i}) \ (v) = 0$ for all v.

So, $(T - \lambda_1)^{n_1} \ (v_1) = 0$ where $v_1 = (\prod_{i \neq 1} (T - \lambda_i)^{n_i}) \ (v)$.

Hence $(T - \lambda_1) \ (\prod_{i \neq 1} (T - \lambda_i)^{n_i}) \ (v) = 0$ for all v, which is a contradiction to the fact that $q(t)$ is the minimum polynomial of T. Thus the minimum polynomial has distinct roots. Hence T is diagonalisable as V can be decomposed as sum of spaces annihilated by $T - \lambda_i$, $1 \leq i \leq s$. Conversely assume that T is diagonalisable. Then there exists a basis $e_1 \ldots e_n$ such that

$$Te_i = \lambda_i e_i, \ 1 \leq i \leq n. \text{ Let } v = \Sigma a_i e_i, \ a_i \neq 0.$$

Then $$(T - \lambda) \ (v) = \Sigma a_i (\lambda_i - \lambda) \ e_i$$

and $$(T - \lambda)^n \ (v) = \sum_i a_i (\lambda_i - \lambda)^n e_i$$

If $(T - \lambda)^n \ (v) = 0$, then $a_i \ (\lambda_i - \lambda)^n = 0$ and hence $a_i \ (\lambda_i - \lambda) = 0$, since $a_i \neq 0$. Thus $(T - \lambda) \ v = 0$.

7. Show that the linear map satisfying $T^2 = T$ is diagonalisable.

Solution: The minimum polynomial of T must divide $t^2 - t = t(t - 1)$. Hence it has distinct roots. Hence T is diagonalisable as V can be decomposed into a sum of subspaces annihilated by T and $T - I$ and each of these spaces is diagonalisable.

8. Let ch $F = 0$ and T satisfy $T^n = 1$. Show that if T has all the characteristic roots in F, then T is diagonalisable.

Solution: Since T has all the characteristic root in F, the minimum polynomial of T is $q\ (t) = \prod_i (t - \lambda_i)^{n_i}$. Now we claim that all these roots are simple. Since $T^n = I$, $q(t)$ divides $t^n - 1$, and if $q(t)$ has a multiple root, so does $t^n - 1$. But $t^n - 1$ cannot have multiple root in F as ch $F = 0$, and there is no common root between $t^n - 1$ and its derivative.

9. Find all possible Jordan forms A for 6×6 matrices with $x^2(1 - x)^2$ as minimum polynomial.

Solution: The possible characteristic polynomials of A have the same irreducible factors x and $(1 - x)$ with atleast square exponents and of degree 6. Hence we have the following 3 cases.

Case (i): Characteristic polynomial is $x^4\ (x - 1)^2$. Jordan form in this case is

$$\begin{pmatrix} 0 & 1 & 0 & 0 & 0 & 0 \\ 0 & 0 & 0 & 0 & 0 & 0 \\ 0 & 0 & 0 & 1 & 0 & 0 \\ 0 & 0 & 0 & 0 & 0 & 0 \\ 0 & 0 & 0 & 0 & 1 & 1 \\ 0 & 0 & 0 & 0 & 0 & 1 \end{pmatrix} \text{ or } \begin{pmatrix} 0 & 1 & 0 & 0 & 0 & 0 \\ 0 & 0 & 0 & 0 & 0 & 0 \\ 0 & 0 & 0 & 0 & 0 & 0 \\ 0 & 0 & 0 & 0 & 0 & 0 \\ 0 & 0 & 0 & 0 & 1 & 1 \\ 0 & 0 & 0 & 0 & 0 & 1 \end{pmatrix}$$

Case (ii): Characteristic polynomial is $x^3\ (x-1)^3$. Jordan from is

$$\begin{pmatrix} 0 & 1 & 0 & 0 & 0 & 0 \\ 0 & 0 & 0 & 0 & 0 & 0 \\ 0 & 0 & 0 & 0 & 0 & 0 \\ 0 & 0 & 0 & 1 & 1 & 0 \\ 0 & 0 & 0 & 0 & 1 & 0 \\ 0 & 0 & 0 & 0 & 0 & 1 \end{pmatrix}$$

Case (iii): Characteristic polynomials is $x^2\ (x-1)^4$. Jordan form is

$$\begin{pmatrix} 0 & 1 & 0 & 0 & 0 & 0 \\ 0 & 0 & 0 & 0 & 0 & 0 \\ 0 & 0 & 1 & 1 & 0 & 0 \\ 0 & 0 & 0 & 1 & 0 & 0 \\ 0 & 0 & 0 & 0 & 1 & 1 \\ 0 & 0 & 0 & 0 & 0 & 1 \end{pmatrix} \text{ or } \begin{pmatrix} 0 & 1 & 0 & 0 & 0 & 0 \\ 0 & 0 & 0 & 0 & 0 & 0 \\ 0 & 0 & 1 & 1 & 0 & 0 \\ 0 & 0 & 0 & 1 & 0 & 0 \\ 0 & 0 & 0 & 0 & 1 & 0 \\ 0 & 0 & 0 & 0 & 0 & 1 \end{pmatrix}$$

10. Find all possible rational canonical forms for 6×6 matrices with $(x-2)$ $(x+2)^3$ as minimum polynomial.

Solution: Since the matrix is of size 6×6 and the minimum polynomial is $(x-2)\ (x+2)^3$ there are three cases for the characteristic polynomial.

Case I: Characteristic polynomial is $(x-2)\ (x+2)^5$. In this case, the rational canonical forms are

$$C(x-2) \oplus C(x+2)^3 \oplus C(x+2)^2$$

or $$C(x-2) \oplus C(x+2)^3 \oplus C(x+2) \oplus C(x+2).$$

Case II: Characteristic polynomial is $(x-2)^2\ (x+2)^4$. In this case, the rational canonical from is

$$C(x-2) \oplus C(x-2) \oplus C(x+2)^3 \oplus C(x+2).$$

Case III: Characteristic polynomial is $(x-2)^3\ (x+2)^3$. In this case, the rational canonical from is

$$C(x-2) \oplus C(x-2) \oplus C(x-2) \oplus C(x+2)^3$$

All these can be written in the block matrix from using the matrix form of $C(q(x))$, the companion matrix of $q\,(x)$.

5.10 INNER PRODUCT SPACES

1. Let V be an inner product space and $d : V \times V \to F$ defined by $d(v, w) = \| v - w \|$. Show that d satisfies the following properties.

(*i*) $d(v, w) \geq 0$ and equal to 0 if, and only if, $v = w$.

(*ii*) $d(v, w) = d\,(w, v)$

(*iii*) $d(v, w') \leq d(v, w) + d(w, w')$.

Solution: (*i*) and (*ii*) are clear.

(*iii*)
$$\begin{aligned} d(v, w') &= \| v - w' \| = \| (v - w) + (w - w') \| \\ &\leq \| v - w \| + \| w - w' \| \\ &= d(v, w) + d\,(w, w') \end{aligned}$$

2. Show that for any $v, v' \in V$,

$$\| v + v' \|^2 + \| v - v' \|^2 = 2(\| v \|^2 + \| v' \|^2)$$

Solution:

$$(v + v', v + v') = \| v + v' \|^2$$

$$(v - v', v - v') = \| v - v' \|^2$$

So $$\| v + v'\|^2 + \| v - v' \|^2 = \| v \|^2 + (v, v') + (v', v)$$

$$+ \| v' \|^2 + \| v \|^2 - (v', v) - (v, v') + \| v' \|^2$$

$$= 2 \| v \|^2 + 2 \| v' \|^2.$$

3. Show that $\| v + v' \|^2 = \| v \|^2 + \| v' \|^2$ if , and only if, v and v' are orthogonal.

Solution:

$$\| v + v'\|^2 = \| v \|^2 + \| v' \|^2 + (v, v') + (v', v)$$

$$= \| v \|^2 + \| v' \|^2$$

if v and v' are orthogonal.

Conversely if $\| v + v' \|^2 = \| v \|^2 + \| v' \|^2$

then $(v, v') + (v', v) = 0$ for all v, v'. In particular, the real part of $(v, v') = 0$. Changing v' to v' to iv', imaginary part of $(v, v') = 0$. So $(v, v') = 0$.

4. Interpret Schwarz inequality geometrically in the inner product space R^2.

Solution: Schwarz inequality states:

$$| a_1b_1 + a_2b_2 | \leq (\sqrt{a_1^2 + a_2^2})(\sqrt{b_1^2 + b_2^2})$$

for $(a_1, a_2), (b_1, b_2) \in \mathbb{R}^2$. If θ is the angle between the vectors (a_1, a_2) and (b_1, b_2) and if they are unit vectors, the inequality states that $| \cos \theta | \leq 1$.

5. Show that the space of all real 2×2 matrices form an inner product space for the product $(A, B) = \text{Tr}\,(AB^t)$.

Solution: (*i*) Tr. $(AA') \geq 0$ and is equal to 0 if, and only if, $A = 0$.

(*ii*) Tr (AA^t) = Tr. (A^tA)

(*iii*) Tr (AB^t) is linear in both A and B. Hence the space is an inner product space.

6. Let $V = R_2\,[t]$ be the space of all polynomials of degree utmost 2 over R. Define an inner product by $(x, y) = \int_0^1 x(t)\, y(t) dt$. Show how to obtain an orthonormal basis form the basis $\{1, t, t^2\}$ of V over R.

Solution: We use the Gram-Schmidt orthonormalisation process starting with basis $\{1, t, t^2\}$. Let $v_1 = 1$, $v_2 = t$ and $v_3 = t^2$.

Then $$e_1 = \frac{v_1}{\| v_1 \|} = 1, \text{ since } \| v_1 \| = 1.$$

$$v_2' = v_2 - (v_2, e_1)\, e_1$$

$$= t - \int_0^1 t\, dt = t - 1/2$$

$$\| v_2' \|^2 = \int_0^1 (t - 1/2)^2 \, dt = 1/2$$

$$e_2 = \frac{v_2'}{\| v_2' \|} = \sqrt{12}\,(t - 1/2)$$

$$v_3' = v_3 - (v_3, e_1)\, e_1 - (v_3, e_2)\, e_2$$

$$= t^2 - \int_0^1 t^2 \, dt - \left(\int_0^1 \sqrt{12}\, t^2\, (t - 1/2)\, dt \right) \sqrt{12}\,(t - 1/2)$$

$$\| v_3' \|^2 = \int_0^1 v_3'(t)\, v_3'(t)\, dt$$

and $e_3 = \dfrac{v_3'}{\| v_3' \|} \cdot \{e_1, e_2, e_3\}$ is an orthonormal basis.

7. Write down the Schwarz inequality for the space given in Problem 5.
Solution: $(A, B) = \text{Tr}\,(AB^t)$
Schwarz inequality gives

$$|\, \text{Tr}\,(AB^t) \,| \le |\, \text{Tr}\,(AA^t) \,|^{1/2} \,|\, \text{Tr}\,(BB^t) \,|^{1/2}.$$

8. Let $V = \mathbb{R}^3$ and let $v_1 = (1, 0, 2)$ and $v_2 = (0, 3, 1)$. Find a vector which is both orthogonal to v_1 and v_2.
Solution: Let (x, y, z) be the required vector. Then $x + 2z = 0$ and $3y + z = 0$. Taking $z = 1$ one such vector is $(-2, -1/3, 1)$.

9. Show that any orthonormal set can be completed to an orthonormal basis.
Solution: Let $\{e_1, \ldots, e_m\}$ be an orthonormal set. Complete it to a basis $\{e_1, \ldots, e_m, v_{m+1}, \ldots, v_n\}$ of V. Perform Gram-Schmidt orthogonalisations process from the element v_{m+1} onwards. We get an orthonormal basis of V.

10. Let $\{v_1, v_2, \ldots, v_n\}$ be an orthonormal set in V. Show that for any $v \in V$, $\sum_{i=1}^{m} |\,(v_i\, v_i)\,|^2 \le \| v \|^2$ (Bessel's inequality). Show further that equality holds if $\{v_1, v_2, \ldots, v_m\}$ is an orthonormal basis of V. (Parseval equality).
Solution: Let $(v, v_1) = a_i$. Then

$$\left\| v - \sum_{i}^{m} a_i\, v_i \right\|^2 = (v - \Sigma\, a_i\, v_i,\; v - \Sigma\, a_i\, v_i)$$

$$= \| v \|^2 - \Sigma a_i\, (v_i, v) - \sum_i \bar{a}_i\, (v, v_i) + \sum_i |\, a_i \,|^2$$

$$= \| v \|^2 - \sum_i a_i\, \bar{a}_i + \sum_i \bar{a}_i\, a_i + \sum_i |\, a_i \,|^2$$

$$= \| v \|^2 - \sum_i |\, a_i \,|^2 \ge 0$$

So $$\sum_{i=1}^{m} |\, a_i \,|^2 \le \| v \|^2$$

i.e., $$\sum_{i=1}^{m} |\,(v, v_i)\,|^2 \le \| v \|^2$$

(*ii*) Suppose $\{v_1, \ldots, v_n\}$ is an orthonormal basis. Write $v = \sum_{i=1}^{m} a_i v_i$. Then

$(v, v_i) = a_i$, $1 \le i \le n$.

$$0 = \left\| v - \sum_{i}^{n} a_i v_i \right\|^2 = \| v \|^2 - \sum_{i=1}^{n} | a_i |^2$$

So $$\| v \|^2 = \sum_{i=1}^{n} | a_i |^2 = \sum_{i=1}^{n} | (v, v_i) |^2 .$$

11. Let $\{v_1, \ldots, v_m\}$ be an orthonormal set in V such that for any $v \in V$, $\Sigma | (v, v_i) |^2$. Show that $\{v_1, \ldots, v_m\}$ is an orthonormal basis.

Solution: Let $a_i = (v, v_i)$. By assumption $\| v \|^2 = \sum_{i=1}^{m} | a_i |^2$ so that

$$\left\| v - \sum_{i=1}^{m} a_i v_i \right\|^2 = \| v \|^2 - \Sigma | a_i |^2 = 0.$$

Hence $v = \sum_{i=1}^{m} a_i v_i$, *i.e.*, $\{v_1 \ldots v_m\}$ generates V. Being an orthonormal set it is linearly independent. Hence it is an orthonormal basis of v.

5.11 UNITARY, HERMITIAN AND ORTHOGONAL MATRICES

1. Determine whether the following matrices are (*a*) unitary (*b*) Hermitian (*c*) Orthogonal (*d*) Skew-Hermitian

(*i*) $\begin{pmatrix} 1 & 2 & 0 \\ 2 & 2 & 2 \\ 0 & 2 & 3 \end{pmatrix}$ (*ii*) $\begin{pmatrix} 1 & 0 & 0 \\ 0 & 1/\sqrt{2} & 1/\sqrt{2} \\ 0 & -1/\sqrt{2} & 1/\sqrt{2} \end{pmatrix}$

(*iii*) $\begin{pmatrix} 0 & -i \\ -i & 0 \end{pmatrix}$.

Solution: (*i*) Hermitian (*ii*) Orthogonal (*iii*) Skew-Hermitian.

2. Show that the determinant of a unitary matrix has absolute value 1.

Solution: Since U is Hermitian $U\bar{U}^t = I = \bar{U}^t U$

So $\quad D(U)\, D(\bar{U}^t) = D(I) = I$

Hence $\quad D(U)\, D(\bar{U}) = 1$ *i.e.*, $D(U)\, \overline{D(U)} = 1$

Hence $D(U)$ has absolute value 1.

3. If A is a unitary matrix, show that (*i*) A^t (*ii*) $\bar{A}$ and (*iii*) A^{-1} are all unitary.

Solution: $A\bar{A}^t = \bar{A}^t A = I$. Let $B = A^t$.

Then $B\bar{B}^t = (A^t)(\bar{A}^t)^t = A^t \bar{A} = I$.

Similarly $\bar{B}^t B = I$. So B is unitary.

(*ii*) Let $B = \bar{A}$. Then $B\bar{B}^t = \bar{A}A^t = \overline{(A\bar{A}^t)} = I$

Similarly $\bar{B}^t B = I$. So B is unitary.

(*iii*) Let $B = A^{-1}$. $B\bar{B}^t = A^{-1}(\bar{A}^{-1})^t = (\bar{A}^t A)^{-1} = I$

Similarly $\bar{B}^t B = I$. Hence B is unitary.

4. Diagonalise the unitary matrix $A = \begin{pmatrix} 0 & -i \\ -i & 0 \end{pmatrix}$.

Solution: The characteristic polynomial of A is

$$D(\lambda I - A) = \begin{vmatrix} \lambda & i \\ i & \lambda \end{vmatrix} = \lambda^2 + 1.$$

So the characteristic roots are $\pm i$. The diagonalisation of A is given by $\begin{pmatrix} i & 0 \\ 0 & -i \end{pmatrix}$.

5. Show that a linear transformation T is Hermitian if, and only if, (Tv, v) is real for all v.

Solution: Since T is Hermitian $(Tv, w) = (v, Tw)$.

In particular, $(Tv, v) = (v, Tv) = \overline{(Tv, v)}$. So (Tv, v) is real.

Conversely assume that (Tv, v) is real for all v.

So $(Tv, v) = \overline{(Tv, v)} = (v, Tv)$ for all v.

In particular $(T(v + w), v + w) = (v + w, T(v + w))$.

This gives $(Tv, w) + (Tw, v) = (v, Tw) + (w, Tv)$.

Change w to iw, cancel i and subtract.

We get $(Tv, w) = (v, Tw)$ for all v, w. Hence T is Hermitian.

6. Show that any $n \times n$ matrix A over C can be written in the form $A = B + iB'$ where B and B' are Hermitian.

Solution:

$$A = \frac{A + \bar{A}^t}{2} + \frac{A - \bar{A}^t}{2}.$$

Let $B = \dfrac{A + \bar{A}^t}{2}$ and $B' = \dfrac{i}{2}(\bar{A}^t - A)$

Then $A = B + iB'$, B, B' are both Hermitian.

7. If A is a Hermitian matrix. Show that $\bar{A}$ and A^t are also Hermitian. If further A is non-singular, show that A^{-1} is also Hermitian.

Solution:

(*i*) Let $B = \bar{A}$

Then $\bar{B} = A = \bar{A}^t = B^t$. So that B is Hermitian.

(*ii*) Let $B = A^t$.

Then $\bar{B} = \bar{A}^t = A = B^t$. So that B is Hermitian.

(*iii*) Let $B = A^{-1}$.

Then $B^t = (A^{-1})^t = (A^t)^{-1} = \bar{A}^{-1} = \bar{B}$. So that B is Hermitian.

8. Let A be a non-zero Hermitian matrix. Show that Tr. $(AA^*) > 0$.

Solution: $A^* = \bar{A}^t$. Tr $(AA^*) = $ Tr $(A\bar{A}^t)$

The diagonal entries of $A\bar{A}^t$ are, respectively,

$$|a_{11}|^2 + |a_{12}|^2 + \ldots + |a_{1n}|^2, |a_{21}|^2 + \ldots + |a_{2n}|^2, \ldots$$

So the Trace $= \sum_{ij} |a_{ij}|^2 > 0$ if at least one $a_{ij} \neq 0$.

9. Show that the sum of two Hermitian matrices is Hermitian and the same is true of the product when they commute with each other.

Solution: Let A, B be Hermitian so that $\bar{A} = A^t$ and $\bar{B} = B^t$.

Then $\overline{A+B} = \bar{A} + \bar{B} = A^t + B^t = (A + B)^t$,

so that $A + B$ is Hermitian. Now assume $AB = BA$.

Then $\overline{AB} = \bar{A}\bar{B} = A^tB^t = (BA)^t = (AB)^t$

so that AB is Hermitian.

10. Show that a triangular Hermitian matrix is diagonal.

Solution: Let A be triangular and Hermitian. If A is lower triangular, A^t is upper triangular.

Since $A = \bar{A}^t$, all of diagonal entries are zero.

11. Show that an orthogonal matrix has determined $\pm$ 1.

Solution: Let A be orthogonal so that $AA^t = I = A^tA$.

Then $D(A)\ D(A^t) = 1$, *i.e.*, $D(A)^2 = 1$. Hence $D(A) = \pm 1$.

12. Show that if A is a diagonal orthogonal matrix, the diagonal elements are all $\pm$ 1.

Solution: Let $A = (d_1 \ldots d_n)$ be diagonal orthogonal.

$A^t = (d_1 \ldots d_n)$. Since $AA^t = I$, $d_i^2 = 1$, for each i, *i.e.*, $d_i = \pm 1$.

13. If A is an orthogonal matrix, show that A^t and A^{-1} are orthogonal.

Solution: $AA^t = I = A^tA$. Let $B = A^t$. Then $BB^t = A^tA = I$.

Similarly $B^tB = I$. Hence B is orthogonal.

Let $B = A^{-1}$. Then $BB^t = A^{-1}(A^{-1})^t = A^{-1}(A^t)^{-1} = (A^tA)^{-1} = I$.

Similarly $B^tB = I$. Hence B is orthogonal.

14. Show that any 2×2 orthogonal matrix is of the form $\begin{pmatrix} \cos\theta & -\sin\theta \\ \sin\theta & \cos\theta \end{pmatrix}$ or $\begin{pmatrix} \cos\theta & \sin\theta \\ \sin\theta & -\cos\theta \end{pmatrix}$ for some θ.

Solution: Let $A = \begin{pmatrix} a & b \\ c & d \end{pmatrix}$ be a 2×2 orthogonal matrix.

Then $\quad a^2 + b^2 = 1 = c^2 + d^2,\ ac + bd = 0.$

Let $a = \cos\theta$ for some θ. Then $b = \pm\sin\theta$. (c, d) is a unit vector orthogonal to (a, c).

Hence if $\quad (a, b) = (\cos\theta, -\sin\theta), (c, d) = (\sin\theta, \cos\theta)$

If $\quad (a, b) = (\cos\theta, \sin\theta)$, then $(c, d) = (\sin\theta, -\cos\theta)$.

This gives the required forms of the 2×2 orthogonal matrix.

15. Reduce to canonical form the orthogonal matrix.

$$A = \begin{pmatrix} 2/2 & -2/3 & 1/3 \\ 2/3 & 1/3 & -2/3 \\ 1/3 & 2/3 & 2/3 \end{pmatrix}$$

Solution: On computing the characteristic polynomial of A, we find that it is $(x-1)(x^2 - 2/3x + 1)$. Then roots of $x^2 - 2/3x + 1$ are $\frac{1}{3} \pm \frac{i2\sqrt{2}}{9}$, *i.e.,* $e^{+i\theta}$, where $\cos\theta = 1/3$ and $\sin\theta = \frac{2\sqrt{2}}{9}$. Thus the orthogonal canonical form of A is

$$\begin{bmatrix} 1 & 0 & 0 \\ 0 & \frac{1}{3} & \frac{-2\sqrt{2}}{9} \\ 0 & \frac{2\sqrt{2}}{9} & \frac{1}{3} \end{bmatrix}.$$

5.12 QUADRATIC FORMS

1. Determine the rank and signature of the following matrices:

(*i*) $\begin{pmatrix} 4 & 5 & -2 \\ 5 & 1 & -2 \\ -2 & -2 & 3 \end{pmatrix}$ $\qquad$ (*ii*) $\begin{pmatrix} 1 & 0 & -1 \\ 0 & 2 & 1 \\ -1 & 1 & 0 \end{pmatrix}$

Solution:

(*i*) $A = \begin{pmatrix} 4 & 5 & -2 \\ 5 & 1 & -2 \\ -2 & -2 & 3 \end{pmatrix}$

The characteristic polynomial of A is

$$\lambda^3 - 8\lambda^2 - 14\lambda + 43.$$

It has, two positive and one negative root and the matrix A is non-singular. Hence rank $A = 3$ and signature is 1.

(*ii*) $A = \begin{pmatrix} 1 & 0 & -1 \\ 0 & 2 & 1 \\ -1 & 1 & 0 \end{pmatrix}$

The characteristic polynomial of A is, $\lambda^3 - 3\lambda^2 + 3$. It has two positive and one negative root. Hence rank $A = 3$ and signature is 1.

2. Reduce to canonical form the quadratic form

$$x_1^2 - 2x_2^2 + 5x_3^2 - 2x_1x_2 + 4x_2x_3 - 16x_3x_1.$$

Solution: The matrix of the form is $A = \begin{pmatrix} 1 & -1 & -8 \\ -1 & -2 & 2 \\ -8 & 2 & 5 \end{pmatrix}$. The characteristic polynomial of A is

$$\lambda^3 - 4\lambda^2 - 76\lambda - 141.$$

It has three negative roots. The canonical form is $-x_1^2 - x_2^2 - x_3^2$.

3. Find the rank and signature of the quadratic form $x_1^2 - 4x_1x_2 + x_2^2$.

Solution: The matrix $A = \begin{pmatrix} 1 & -2 & 0 \\ -2 & 1 & 0 \\ 0 & 0 & 0 \end{pmatrix}$ has rank 2 and characteristic polynomial $\lambda^3 - 2\lambda^2 - 3\lambda$.

The roots are $\lambda = 0, -1$ and 3. Hence signature is 0.

4. Investigate whether the following quadratic forms are (*a*) positive definite (*b*) negative definite (*c*) indefinite.

(*i*) $x_1^2 - 2x_1x_2 + x_2^2 + 4x_2x_3 + 5x_3^2$

(*ii*) $5x_1x_2 - x_3^2$

(*iii*) $2x_1^2 + 3x_2^2 + x_3^2$.

Solution:

(*i*) $A = \begin{pmatrix} 1 & -1 & 0 \\ -1 & 1 & 2 \\ 0 & 2 & 5 \end{pmatrix}$ has characteristic polynomial $\lambda^3 - 7\lambda^2 + 6\lambda + 4$. It has two positive and one negative root. Hence it is indefinite.

(*ii*) $A = \begin{pmatrix} 0 & 5/2 & 0 \\ 5/2 & 0 & 0 \\ 0 & 0 & -1 \end{pmatrix}$ has characteristic polynomial $\lambda^3 + \lambda^2 - 25/4\lambda - 25/4 = (\lambda + 1)(\lambda^2 - 25/4)$. The characteristic roots are $-1, 5/2, -5/2$. Hence it is indefinite.

(*iii*) $\begin{pmatrix} 2 & 0 & 0 \\ 0 & 3 & 0 \\ 0 & 0 & 1 \end{pmatrix}$ has characteristic roots 2, 3, 1. Hence it is positive definite.

5. Show that the real quadratic form $ax_1^2 + bx_1x_2 + cx_2^2$ in two variables $x_1 x_2$, is positive definite if, and only if, $a > 0$ and $b^2 - 4ac < 0$.

Solution: $A = \begin{pmatrix} a & a/2 \\ b/2 & c \end{pmatrix}$ has characteristic polynomial

$$\lambda^2 - (a + c)\lambda + \frac{4ac - b^2}{4}.$$

If λ_1, λ_2 are the characteristic roots

$$\lambda_1 + \lambda_2 = a + c \text{ and } \lambda_1\lambda_2 = \frac{4ac - b^2}{4}.$$

If $b^2 - 4ac < 0$, $\lambda_1\lambda_2 > 0$. If $a > 0$, $4ac > b^2 > 0$ and hence $c > 0$. Hence $\lambda_1 + \lambda_2 > 0$. Hence both λ_1, λ_2 are positive. Converse is proved by reversing the steps. Hence the form is positive definite if, only if, $a > 0$ and $b^2 < 4ac$.

6. A quadratic form is said to be positive semi-definite if all the characteristic roots are non-negative. Show that a positive semi-definite form whose matrix is non-singular is positive definite.

Solution: Since the form is positive semi-definite all the characteristic roots $\lambda_i \geq 0$. Since it is non-singular, no λ_i can be zero.

Hence $\lambda_i > 0$ for all *i, i.e.,* the form is positive definite.

7. Show that a positive definite form cannot be congruent to an indefinite form.

Solution: A positive definite form has all characteristic roots positive. Since any congruent form also has the same characteristic roots, it cannot be indefinite.

8. Reduce the following pairs of quadratic forms simultaneously to normal form. $4x_1^2 + x_2^2 + 9x_3^2$; $x_1x_2 + x_2x_3 + x_3x_1$.

Solution: Let $A = \begin{pmatrix} 4 & 0 & 0 \\ 0 & 1 & 0 \\ 0 & 0 & 9 \end{pmatrix}$ and $B = \begin{pmatrix} 0 & 1/2 & 1/2 \\ 1/2 & 0 & 1/2 \\ 1/2 & 1/2 & 0 \end{pmatrix}$.

A is positive definite as the characteristic roots of *A* are 4, 1, 9 all positive.

Under simultaneous reduction, the first form reduces to $x_1^2 + x_2^2 + x_3^2$ and the second form to $\lambda_1 x_1^2 + \lambda_2 x_2^2 + \lambda_3 x_3^2$, where $\lambda_1, \lambda_2, \lambda_3$, are the roots $D(\lambda A - B) = 0$.

Now $$\lambda A - B = \begin{pmatrix} 4\lambda & -1/2 & -1/2 \\ -1/2 & \lambda & -1/2 \\ -1/2 & -1/2 & 9\lambda \end{pmatrix}.$$

so that $D(\lambda A - B) = 0$ gives the equation

$$144\lambda^3 - 14\lambda - 1 = 0.$$

Substituting the roots of this equation in the above, we get the normal form for the second.

9. Which of the following forms are congruent to each other? Why?

(i) $x_1^2 + x_2^2 + x_3^2 + x_1x_2$, *(ii)* $x_1x_2 + x_2x_3 + x_3x_1$

(iii) $x_1^2 + 2x_2^2 + 2x_3^2 + 3x_1x_2$.

Solution:

(i) $A = \begin{pmatrix} 1 & 1/2 & 0 \\ 1/2 & 1 & 0 \\ 0 & 0 & 1 \end{pmatrix}$ has characteristic polynomial

$$\lambda - 1\,(\lambda^2 - 2\lambda + 3/4) = \lambda - 1\,(\lambda - 3/2)\,(\lambda - 1/2)$$

All the 3 roots are positive. Hence it has rank 3 and signature 3.

(ii) $A = \begin{pmatrix} 0 & 1/2 & 1/2 \\ 1/2 & 0 & 1/2 \\ 1/2 & 1/2 & 0 \end{pmatrix}$ has characteristic polynomial.

$$\lambda^3 - \frac{3\lambda}{4} - \frac{1}{4}.$$

It has $\lambda = 1$ as a root. Since the product of the roots is positive and the sum of the other two roots is -1, the other two roots are negative. Hence it has rank 3 and signature -1.

(iii) $A = \begin{pmatrix} 1 & 3/2 & 0 \\ 3/2 & 2 & 0 \\ 0 & 0 & 2 \end{pmatrix}$ has characteristic polynomial $(\lambda - 2)(\lambda^2 - 3\lambda - 1/4)$.

It has two positive and one negative root. Hence the rank is 3 and signature is 1. Since all the three forms have different signatures, no two forms are congruent to each other.

10. Find all quadratic forms Q in two variables of rank 2 which are in the normal form. Which conics do the equations $Q(x) = 1$ represent?

Solution: Assume rank Q = 2. The normal forms are $x_1 + x_2^2$, $x_1^2 - x_2^2$ and $-x_1^2 - x_2^2$. The equation $x_1^2 + x_2^2 = 1$ represents a circle, $x_1^2 - x_2^2 = 1$ represents a rectangular hyperbola and $-x_1^2 - x_2^2 = 1$ is the empty conic.

11. Which following symmetric matrices are positive definite? Why?

(*i*) $\begin{pmatrix} 0 & -1 \\ -1 & 0 \end{pmatrix}$ (*ii*) $\begin{pmatrix} 3 & 3 & 1 \\ 3 & 0 & -1 \\ 1 & -1 & 1 \end{pmatrix}$

Solution: (*i*) $A = \begin{pmatrix} 0 & -1 \\ -1 & 0 \end{pmatrix}$ has characteristic polynomial

$$\lambda^2 - 1 = (\lambda + 1)(\lambda - 1).$$

Since one root is negative, the matrix is not positive definite.

(*ii*) $A = \begin{pmatrix} 3 & 3 & 1 \\ 3 & 0 & -1 \\ 1 & -1 & 2 \end{pmatrix}$ has characteristic polynomial

$$\lambda^3 - 5\lambda^2 - 5\lambda + 27.$$

It has two positive and one negative root. Hence the matrix is not positive definite.

12. Show that a symmetric matrix A is positive definite if, and only if, $A = C^t C$ for some non-singular c.

Solution: Assume $A = C^t C$, C non-singular. Then A is congruent to the identity matrix. Since the identity matrix is positive definite, A is positive definite.

Conversely assume that A is positive definite. Then A is congruent to I and hence $A = C^tIC = C^tC$ for some non-singular C.

Then $H = Z_{q'_1} \oplus \ldots \oplus Z_{q'_k}$ is the required subgroup.

13. Find all abelian groups of order 90,000.

Solution: $90{,}000 = 2^4 \times 3^2 \times 5^4$

So the number of abelian groups of order 90,000 are

$$p(4) \times p(2) \times p(4) = 5 \times 2 \times 5 = 50.$$

These can the enumerated as the groups

$$Z_{2^4} \times Z_{3^2} \times Z_{5^4},\ Z_{2^3} \times Z_2 \times Z_{3^2} \times Z_{5^4},\ Z_{2^2} \times Z_2 \times Z_2 \times Z_{3^2} \times Z_{5^4},$$

$Z_{2^2} \times Z_{2^2} \times Z_{3^2} \times Z_{5^4}, \times Z_2 \times Z_2 \times Z_2 \times Z_2 \times Z_{3^2} \times Z_{5^4}$ and so on.

Index

F

G

I

J

K

L

Q

R

S

T

U

V

Z